AF686882

AN EXPLORATION ON EXTRATERRESTRIAL LIFE

An Exploration on Extraterrestrial Life

THOMAS FRANDSEN

Copyright © 2025 by Thomas Frandsen
All rights reserved. No part of this book may be reproduced in any manner whatsoever without written permission except in the case of brief quotations embodied in critical articles and reviews.
First Printing, 2025

CONTENTS

PREFACE

The question of whether we are alone in the universe has captivated humanity for millennia. From ancient myths to modern science fiction, the possibility of extraterrestrial life has fueled our imaginations and inspired countless works of art and scientific inquiry. Recent developments, including a renewed interest in UFO phenomena and advancements in astrobiology, have brought this question into sharper focus. This book is not an attempt to provide definitive answers; rather, it aims to explore the vast landscape of scientific evidence and reasoned speculation surrounding the search for extraterrestrial life. Using the extraordinary biodiversity of our own planet as a guide, we will examine the potential for life to flourish under a wide range of conditions. We will delve into the intricacies of extremophile biology, the chemistry of life, and the vastness of the cosmos, exploring the potential for life to exist in environments previously deemed uninhabitable. We will examine credible research, evaluate existing data, and address some of the most intriguing and challenging questions related to potential alien contact. While acknowledging the unknowns, this book offers a comprehensive exploration of the exciting possibilities that lie ahead and challenges us to consider the implications of a universe teeming with life. The journey will involve exploring the scientific method, accepting the existence of unknowns, and embracing the potential for unexpected discoveries. It is a journey of reasoned speculation and rigorous scientific analysis, inviting the reader to ponder the mysteries of existence alongside us.

INTRODUCTION

For decades, the topic of extraterrestrial life remained largely relegated to the realm of science fiction. However, recent shifts in global discourse, including a reassessment of UFO phenomena by various governments worldwide and the accelerating advancements in astrobiology, have significantly altered the landscape. This book embarks on a comprehensive exploration of this evolving narrative, moving beyond sensationalism to engage in a rigorous scientific investigation. We will investigate the potential for life beyond Earth using a multifaceted approach, drawing upon the incredible diversity of life forms found here on our own planet to guide our understanding of what might exist elsewhere. We will explore the concept of extremophiles – organisms thriving in extreme environments that challenge our anthropocentric notions of habitability. By examining their adaptations and the underlying chemical principles supporting life, we can begin to imagine the potential for life in vastly different cosmic environments. This exploration will take us through the vast expanse of the cosmos, highlighting the statistical probability of habitable planets existing beyond our own solar system. We will discuss advancements in telescope technology, spacecraft exploration, and data analysis techniques that are revolutionizing our ability to detect potential biosignatures. However, this journey will also involve grappling with the challenges and limitations inherent in the search for extraterrestrial life. We'll evaluate the Fermi Paradox, weighing the seemingly contradictory evidence of a statistically high probability of extraterrestrial life against the apparent lack of confirmed contact. The book will also carefully address the perplexing phenomenon of reported alien abduction accounts, assessing them within a framework of scientific rigor while acknowledging the difficulties of definitively analyzing such accounts. Ultimately, this book strives to be a reasoned and balanced exploration of the evidence and speculation surrounding extraterrestrial life, encouraging a critical and open-minded approach to this momentous scientific and philosophical pursuit. The ultimate goal is to foster a deeper understanding of our place in the universe and prepare for the incredible possibilities that lie ahead.

| 1 |

Renewed Interest in Extraterrestrial Life

For decades, the subject of Unidentified Aerial Phenomena (UAPs), often colloquially referred to as UFOs, occupied a fringe space in public discourse. Dismissal, ridicule, and conspiracy theories were the dominant narratives, fueled by a lack of transparency from governments and a preponderance of unreliable or sensationalized accounts. The very mention of UFOs often invoked images of grainy photographs, shaky home videos, and fantastical claims of alien abductions, firmly relegating the topic to the realm of science fiction for many. This entrenched skepticism, however, is undergoing a significant shift. A subtle but palpable change is occurring, marked by a growing willingness from government agencies and scientific institutions to engage with the phenomenon in a more serious and systematic manner.

This change isn't solely a result of a sudden influx of high-quality evidence. While some compelling UAP sightings have undoubtedly contributed, the shift is also a product of evolving societal attitudes, advances in technology, and a growing recognition of the limitations of our current understanding of the atmospheric and aerospace domains. For years, the official position of many governments was one of outright denial or dismissive indifference. Reports of unexplained aerial objects were often attributed to misidentified aircraft, weather phenomena, or outright hoaxes. This blanket dismissal, while convenient, served to stifle open investigation and fueled public distrust.

The tide began to turn subtly at first. In recent years, we've witnessed a slow but steady release of declassified documents from various government agencies. These documents, while often redacted and incomplete, offer glimpses into past investigations of UAPs, revealing a level of seriousness previously unseen. The sheer volume of these declassified reports, spanning decades, collectively suggests a pattern of unexplained aerial observations that cannot be so easily dismissed as simple errors or hoaxes. For example, the release of the Pentagon's Preliminary Assessment: Unidentified Aerial Phenomena in 2021, a report prepared by the Unidentified Aerial Phenomena Task Force (UAPTF), marked a significant turning point. This report, while cautious in its conclusions, acknowledged the existence of UAPs that defied easy explanation and stressed the need for further investigation.

The report, however, wasn't a sudden revelation of definitive proof of extraterrestrial life. Rather, it was a recognition of a phenomenon that deserves rigorous scientific scrutiny. This shift in tone is crucial. Instead of outright dismissal, the report adopted an approach of careful analysis, acknowledging the limitations of current understanding and the need for more sophisticated data collection and analysis methods. This marks a departure from the previous strategy of simply brushing aside reports as unsubstantiated claims.

This change in government attitude is mirrored by a shift in public perception. The stigma surrounding UAPs is gradually fading, replaced by a growing curiosity and a willingness to engage with the topic in a more nuanced way. Increased media coverage, driven by the release of government documents and the growing interest from credible scientists and researchers, has helped to normalize the discussion. This has allowed for a more open and informed dialogue, moving away from sensationalism and towards a more rational and scientific approach.

Several factors have contributed to this shift in public perception. The improved quality of eyewitness accounts, often accompanied by video and other supporting evidence, plays a role. While these accounts alone may not provide irrefutable proof, they do offer data points that deserve consideration. Moreover, the ubiquity of camera-equipped smartphones and drones has greatly increased the availability of visual documentation, leading to a surge in reported

UAP sightings. Many of these reports come from highly credible sources, such as military pilots or air traffic controllers, lending greater weight to the observed phenomena.

Additionally, the increasing availability of information online has allowed for a more informed discussion. The internet has created a space for sharing information, analyzing data, and fostering collaboration among researchers and enthusiasts alike. While this increased accessibility also facilitates the spread of misinformation, it has simultaneously fostered a more robust and critical assessment of available evidence. This level of informed public engagement demands that governmental and scientific institutions respond with transparency and engagement, rather than continued dismissiveness.

The changing narrative is further fueled by the growing recognition within the scientific community that our understanding of the universe is still fundamentally limited. The sheer scale of the cosmos, the vast number of potentially habitable planets, and the incredible adaptability of life on Earth all suggest the possibility of life beyond our own planet. The discovery of extremophiles – organisms thriving in extreme environments on Earth – has significantly expanded our understanding of what conditions can support life. These extremophiles, found in deep-sea hydrothermal vents, acidic hot springs, and even within nuclear reactors, challenge our anthropocentric views of habitability, suggesting that life may exist in environments previously considered uninhabitable.

This expanded understanding of life's adaptability, coupled with the ongoing discovery of exoplanets – planets orbiting stars other than our sun – strengthens the case for the potential existence of extraterrestrial life. Thousands of exoplanets have been discovered, with many residing within the habitable zones of their respective stars, meaning that they have the potential to support liquid water on their surface. The discovery of these planets, many with characteristics similar to Earth, dramatically increases the probability of life existing elsewhere in the universe.

Furthermore, the limitations of our current scientific instruments and methods must be acknowledged. Our ability to detect biosignatures – chemical or physical indicators of life – beyond Earth is still in its infancy. The distances involved are immense, the signals weak, and our understanding of extraterrestrial life forms remains largely speculative. We are essentially looking for something we don't fully understand, using tools that may not be entirely suitable for the task. This inherent uncertainty necessitates a shift towards a more open and explorative approach.

The ongoing evolution of technology also plays a crucial role. Advancements in telescope technology, spacecraft design, and data analysis techniques are continually improving our ability to detect and characterize exoplanets and search for biosignatures. As these technologies improve, our understanding of the universe and our chances of detecting extraterrestrial life are likely to increase exponentially. This ongoing technological evolution necessitates a continued commitment to research and development, further supporting the shift towards a more serious investigation of UAPs.

In conclusion, the changing narrative on UFOs is not simply about the possibility of alien visitations. It's about a fundamental shift in how we approach the unknown, a recognition of the limitations of our current understanding, and a growing willingness to engage with challenging questions in a more open and scientific manner. The increased transparency from governments, the growing public interest, and the ongoing advancements in science and technology are collectively contributing to a paradigm shift, one that could redefine our understanding of life, the universe, and our place within it. The journey has just begun, and the coming decades promise to be pivotal in unraveling the mysteries surrounding UAPs and the profound implications of their existence.

The renewed interest in extraterrestrial life, fueled by a reassessment of UAPs and a growing acceptance of the vastness and potential of the cosmos, necessitates a clear definition of the scope of our search. This book is not a compendium of every alleged alien encounter or a treatise on fringe theories. Instead, it aims to navigate the complex landscape of potential extraterrestrial contact through the lens of rigorous scientific inquiry. This requires establishing specific criteria for evaluating evidence and distinguishing between scientifically plausible investigations and unsubstantiated claims.

One of the primary challenges lies in separating credible data from the overwhelming volume of anecdotal evidence, misinterpretations, and outright hoaxes that have historically plagued the field of UAP research. Many reports, while potentially intriguing, lack the verifiable evidence necessary for robust scientific analysis. Grainy photographs, blurry videos, and subjective eyewitness accounts, though sometimes compelling in their narrative, often fail to meet the standards of reproducibility and falsifiability fundamental to the scientific method. While such accounts might spark further investigation, they cannot, on their own, constitute compelling evidence for extraterrestrial contact. This book will critically assess such accounts, acknowledging their existence while carefully evaluating their scientific validity.

Our focus, therefore, will be on evidence that can be subjected to rigorous scientific scrutiny. This includes data that is:

Verifiable: Evidence must be capable of being independently confirmed through multiple sources and methods. This means that data should not rely solely on a single eyewitness account or a single piece of technological evidence. Instead, corroboration from multiple independent sources is crucial. For example, radar data corroborating visual sightings, or multiple independent witness testimonies aligned with physical evidence, would be considered more reliable than a solitary, uncorroborated report.

Reproducible: Scientific findings must be reproducible; other researchers should be able to repeat the same observations or experiments under similar conditions and achieve comparable results. This principle is crucial for validating any claim of extraterrestrial contact. If a phenomenon cannot be replicated or independently verified, its scientific validity is questionable.

Falsifiable: A scientific hypothesis must be falsifiable, meaning that there must be a potential way to prove it wrong. This means that evidence presented must be susceptible to testing and potential refutation. Claims that cannot be tested or challenged through scientific methods are inherently unscientific and will be excluded from our detailed analysis.

Quantifiable: Whenever possible, evidence should be quantifiable, meaning it can be measured and expressed numerically. This allows for objective analysis and comparison. For instance, precise measurements of an object's speed, trajectory, and spectral signature are far more valuable than vague descriptions. The use of quantitative data, rather than qualitative descriptions alone, is crucial for the rigorous scientific evaluation of potential evidence.

Using these criteria, we will examine existing data sets, carefully assessing their quality, reliability, and potential biases. This approach necessitates a thorough review of the methodology used in data collection, analysis, and interpretation. Any potential limitations or biases inherent in the data or methods employed will be openly acknowledged and discussed.

The book will also prioritize the examination of physical evidence. While eyewitness testimony can be valuable, physical remnants, such as materials with unusual compositions, or technologies exhibiting inexplicable functionality, hold the potential to provide more objective and verifiable evidence. The analysis of such materials, using sophisticated techniques like spectroscopic analysis and material characterization, can offer critical insights into their origin and nature, potentially revealing evidence that transcends terrestrial explanations.

We will also incorporate insights from related fields, such as astrophysics, planetary science, and biochemistry. Understanding the conditions necessary for life to arise and thrive, both on Earth and potentially elsewhere in the universe, is essential for evaluating claims of extraterrestrial contact. The book will draw upon current scientific knowledge concerning the prevalence of exoplanets, the potential habitability of other celestial bodies, and the remarkable diversity of life found on our own planet.

This book will not shy away from exploring the implications of extraterrestrial contact, but it will do so in a responsible and scientifically grounded manner. While acknowledging the profound potential impact of such a discovery on human civilization, we will focus on analyzing the evidence critically and objectively, without resorting to speculation or sensationalism. The goal is not to provide definitive answers, but rather to foster a more informed and nuanced understanding of the search for extraterrestrial life.

This requires a conscious effort to avoid falling into common pitfalls. One such pitfall is confirmation bias, the tendency to favor information that confirms pre-existing beliefs. We will strive for objectivity, acknowledging and addressing potential biases in the data and our own interpretations. Another significant challenge lies in distinguishing genuine anomalies from natural phenomena that are currently poorly understood. A thorough analysis must differentiate between potentially extraordinary occurrences and simply unusual atmospheric or astronomical phenomena.

Furthermore, the book will address the historical context surrounding UAP investigations, highlighting both the progress made and the limitations encountered. This includes acknowledging past governmental secrecy and skepticism, as well as the impact of media portrayals and public perception. A historical perspective provides valuable context for understanding the current shift towards more open and transparent investigations.

This thorough and methodical approach, focused on verifiable, reproducible, falsifiable, and quantifiable evidence, is crucial for navigating the complex and often controversial subject of extraterrestrial life. By adhering to rigorous scientific standards, we aim to contribute to a more informed and evidence-based discussion, moving beyond speculation and focusing on the real possibility of life beyond Earth. The ultimate goal is not to definitively prove or disprove the existence of extraterrestrial life, but rather to systematically assess the available evidence and advance our scientific understanding of this profound question. The journey of exploration and discovery continues, and through rigorous scientific analysis, we strive to illuminate the path forward. The next chapters will explore specific examples of potential evidence, meticulously analyzing their validity and implications within the framework established in this section.

Our current understanding of the universe, vast as it may seem, remains profoundly limited. While advancements in astronomy, astrophysics, and related fields have dramatically expanded our knowledge of the cosmos, the sheer scale and complexity of the universe continue to present insurmountable challenges to a complete understanding of the prevalence of life beyond Earth. This section explores some of the key limitations that currently hinder our ability to definitively assess the potential for extraterrestrial life.

One major limitation lies in the nature of our observational capabilities. Even with the most powerful telescopes currently available, our ability to directly observe exoplanets, especially those orbiting distant stars, remains severely restricted. The immense distances involved, coupled with the relative faintness of exoplanets compared to their host stars, makes direct imaging extremely difficult. We often rely on indirect methods, such as transit photometry (detecting dips in a star's brightness as a planet passes in front of it) or radial velocity measurements (detecting the subtle wobble in a star's motion caused by the gravitational pull of an orbiting planet), to infer the existence and properties of exoplanets. These techniques are powerful but have inherent limitations. For example, transit photometry is biased towards planets that transit directly in front of their star from our perspective, and radial velocity measurements are more sensitive to larger planets closer to their stars. This means that we may be missing many smaller, potentially habitable planets located further from their star.

Furthermore, our ability to characterize the atmospheres of exoplanets is still in its infancy. While we have made some progress in detecting atmospheric components through spectroscopy – analyzing the light that passes through an exoplanet's atmosphere – the signals are often weak and difficult to interpret. The technology required for detailed atmospheric analysis is demanding and requires substantial advancements to provide definitive evidence of biomarkers, such as oxygen, methane, or water vapor, that could indicate the presence of life. The detection of such biomarkers is crucial but not conclusive; their presence could be due to non-biological processes. Conversely, the absence of detectable biomarkers doesn't necessarily rule out life, as it might be present but undetectable with current technology or might exist in forms that produce different biosignatures.

Beyond the limitations of observation, our understanding of the conditions necessary for life to arise and thrive is still incomplete. While we know that liquid water, a suitable energy source, and essential chemical elements are necessary for life as we know it, the precise combination of factors required for life's emergence remains unknown. The possibility of life forms fundamentally different from those on Earth, with different biochemistry and metabolic processes, poses a significant challenge to our search. Our current search strategies are heavily biased towards life as we know it, potentially overlooking life forms that might exist under vastly different conditions. We are essentially searching for a needle in a cosmic haystack, and the needle itself may be far more diverse and unpredictable than we currently imagine.

Another crucial limitation relates to the sheer vastness of the universe. Even if life is relatively common, the distances between potentially habitable planets are immense, making direct exploration a daunting, if not presently insurmountable, task. Interstellar travel, as currently conceived, faces technological challenges that appear insurmountable at this stage of our technological development. The time required for interstellar voyages using our current propulsion systems would be measured in centuries, if not millennia. The possibility of faster-than-light travel remains firmly in the realm of science fiction, and even if it were achievable, the energetic requirements and potential risks remain to be fully understood.

Our current understanding of the very origins of life on Earth also presents significant limitations. Despite extensive research, the precise mechanisms that led to the emergence of life from non-living matter remain largely unclear. Understanding abiogenesis – the transition from non-living matter to living organisms – is vital for understanding the potential for life to arise on other planets. If abiogenesis is a rare event, the probability of life existing elsewhere might be significantly lower than if it occurs relatively easily under appropriate conditions. Unraveling the mysteries of abiogenesis requires further breakthroughs in fields like biochemistry, molecular biology, and geochemistry.

Furthermore, the potential for panspermia, the hypothesis that life can spread throughout the universe through the transfer of microbial life between celestial bodies, introduces another layer of complexity. If panspermia is a significant factor in the distribution of life, it could explain the apparent similarity of life forms across the universe, but it also raises challenges in understanding the origin of life itself, potentially shifting the focus from abiogenesis on a single planet to a broader cosmic context. The mechanisms and likelihood of successful panspermia are currently not fully understood.

Moreover, it is important to acknowledge the potential for unknown factors that might influence the search for extraterrestrial life. We might be overlooking crucial aspects of astrobiology or planetary science that could significantly affect the prevalence of life beyond Earth. The universe might contain phenomena or processes that we are currently unable to conceive, potentially changing our understanding of habitability and life itself. Our current scientific models are based on our current knowledge, and new discoveries might require revisions and adjustments to our hypotheses about extraterrestrial life.

In addition to the scientific limitations, the search for extraterrestrial life is also constrained by technological and logistical challenges. The cost of space exploration is substantial, and funding priorities often favor missions with more immediate or tangible returns. The search for extraterrestrial life is a long-term endeavor that requires sustained investment and international collaboration. The development of advanced technologies, such as more powerful telescopes, sensitive instruments for atmospheric analysis, and advanced propulsion systems, is vital but requires significant research and development efforts. This demands a commitment to long-term scientific goals, sometimes at the expense of shorter-term objectives.

Finally, the social and cultural implications of discovering extraterrestrial life must also be considered. The societal impact of such a discovery would be profound, potentially altering our understanding of humanity's place in the universe and raising ethical and philosophical questions about our relationship with any discovered life forms. The preparation for such an eventuality requires careful consideration of the potential social, political, and cultural ramifications. It calls for open dialogue and international cooperation to ensure that any discovery is handled responsibly and ethically.

In conclusion, while the renewed interest in extraterrestrial life is driven by a combination of scientific advancements, reassessment of UAP data, and a growing awareness of the universe's vastness, the current state of our scientific knowledge presents substantial limitations. Our observational capabilities, incomplete understanding of abiogenesis and the conditions for life, the vast distances involved in interstellar travel, and the possibility of unknown factors all contribute to the challenges in definitively assessing the potential for extraterrestrial life. Overcoming these limitations requires continuous advancements in technology, fundamental research in astrobiology and related fields, and international collaboration to pursue the long-term goal of uncovering the secrets of the cosmos. The quest to find life beyond Earth is an ongoing journey of exploration and discovery, and the challenges we face serve as a reminder of the incredible complexity and mystery of the universe. The scientific endeavor, driven by curiosity and a commitment to truth, will continue to unravel the mysteries of extraterrestrial life, even if the answers may take time to surface.

The sheer scale of the universe often defies our comprehension. We grapple with astronomical distances expressed in light-years, a unit of measurement that encapsulates the vastness in a way that's difficult to truly grasp. A light-year represents the distance light travels in a year – approximately 5.88 trillion miles. To put this into perspective, imagine driving a car at a constant speed of 60 mph. It would take you over 100 billion years to travel just one light-year, far exceeding the age of the universe itself.

This vastness, however, doesn't diminish the possibility of life beyond Earth; it arguably enhances it. The Milky Way galaxy alone is estimated to contain hundreds of billions of stars, each potentially harboring its own planetary system. Many of these stars are similar to our Sun, and the Kepler space telescope and subsequent missions have revealed the abundance of exoplanets – planets orbiting stars other than our Sun – many of which reside within the habitable zones of their stars, the region where liquid water could potentially exist on a planet's surface.

Consider the sheer number of galaxies in the observable universe, estimated to be in the hundreds of billions, each containing its own multitude of stars and planets. The probability of Earth being the sole bearer of life within this cosmic expanse seems vanishingly small, bordering on statistically improbable. The sheer number of potential habitable environments suggests a much higher likelihood of life existing elsewhere.

To illustrate the scale, consider a single grain of sand. Imagine that grain of sand represents a single star in the Milky Way galaxy. Now imagine a beach extending for miles, representing the entire galaxy. Each grain of sand on that beach represents a star with a potential planetary system. Then, visualize countless other beaches, each representing other galaxies, stretching across an unimaginable landscape. This is just a feeble attempt to grasp the sheer scale of the cosmos.

The improbability of life's exclusivity to Earth is further reinforced by the remarkable biodiversity found on our own planet. Life on Earth has adapted to an astounding array of environments, from the crushing depths of the ocean to the scorching heat of volcanic vents, from the frigid Antarctic to the arid deserts. Extremophiles, organisms that thrive in extreme conditions, demonstrate the remarkable adaptability of life and hint at the potential for life to flourish in environments far different from our own. If life could emerge and diversify so dramatically on Earth, under such varied and often harsh conditions, it's reasonable to speculate that similar processes could have occurred on countless other planets across the universe, leading to the evolution of diverse extraterrestrial life forms.

This concept is further underscored by the ongoing discovery of exoplanets with characteristics potentially conducive to life. While we haven't yet found definitive proof of extraterrestrial life, the discovery of exoplanets in habitable zones around other stars significantly strengthens the argument for its possibility. The detection of water vapor or other biosignatures in the atmospheres of some exoplanets adds another layer of intrigue, although more research is needed to ascertain the significance of such findings. The ongoing search for exoplanets and the development of increasingly sophisticated techniques for detecting biosignatures are steadily pushing the boundaries of our knowledge and increasing our chances of discovering evidence of life beyond Earth.

Beyond the mere existence of potentially habitable planets, the universe's age also plays a significant role in considering the probability of life elsewhere. The universe is estimated to be around 13.8 billion years old, giving ample time for life to emerge and evolve on other planets. Even if the emergence of life is a relatively rare event, the sheer span of cosmic time provides ample opportunity for it to have happened multiple times, potentially on countless planets across the vast expanse of the universe. This vast timescale dramatically increases the likelihood of life existing beyond Earth.

The possibility of life fundamentally different from life on Earth further expands the scope of the search. Our terrestrial perspective often biases our search towards life forms similar to those we know, using carbon-based biochemistry and liquid water as essential components. However, the universe might harbor life forms based on alternative biochemistries, utilizing different solvents, energy sources, or even fundamental building blocks. Exploring this possibility requires shifting our perspectives beyond terrestrial biology and embracing the potential for vastly different forms of life. This necessitates exploring diverse environments and searching for biosignatures beyond those anticipated based on our current understanding of life.

The immense scale of the universe, coupled with the vast number of stars and planets, the adaptability of life as demonstrated on Earth, the discovery of exoplanets within habitable zones, the immense age of the universe, and the potential for diverse life forms all converge to paint a picture of a cosmos that is potentially teeming with life. While we have yet to find definitive proof, the circumstantial evidence is compelling, suggesting that we are likely not alone. The ongoing exploration of the cosmos is a testament to humanity's enduring curiosity and the pursuit of answers to some of the most profound questions we can ask – are we alone? The sheer improbability of a solitary, living planet in this vast expanse suggests the answer might be a resounding no. The ongoing search, therefore, is not merely a scientific endeavor, but a quest for a deeper understanding of our place within this grand cosmic tapestry. The more we learn, the more likely it becomes that the universe is teeming with the wonders of life in ways beyond our current imagination.

The preceding chapters have established a compelling case for the plausibility of extraterrestrial life, drawing upon the sheer scale of the universe, the remarkable biodiversity of Earth, and the ongoing discovery of exoplanets within habitable zones. However, the question of extraterrestrial life is not simply a matter of astronomical probabilities; it requires a rigorous scientific approach, grounded in empirical evidence and informed by the latest advancements in astrobiology, planetary science, and related fields. This section outlines the methodology employed throughout this book and highlights the delicate balance between open-mindedness and scientific rigor that is essential for navigating the complexities of this fascinating topic.

One of the primary challenges in the search for extraterrestrial life lies in defining what we are actually looking for. Our current understanding of life is fundamentally based on terrestrial biology – the study of life on Earth. We know life as we know it requires liquid water, a carbon-based biochemistry, and a stable energy source.

However, it's crucial to acknowledge that this Earth-centric perspective might significantly limit our ability to detect life that is fundamentally different. The universe's vastness suggests the possibility of life forms that utilize alternative biochemistries, perhaps based on silicon, ammonia, or other elements. They might thrive in environments that would be lethal to terrestrial organisms, utilizing different energy sources and possessing unique metabolic pathways. Therefore, our search strategy cannot be overly constrained by our current understanding of life's essential requirements.

This book embraces a multi-faceted approach, integrating various scientific disciplines to explore the broader question of extraterrestrial life. We will examine the geological and atmospheric conditions necessary for habitability, drawing on insights from planetary science and geochemistry. We will delve into the possibilities of different biosignatures—chemical or physical indicators of life—that could be detected remotely, through telescopic observations or robotic missions to other planets or moons. This will involve a critical assessment of existing data, acknowledging limitations and uncertainties in current detection methods. Furthermore, the study will explore the potential of different types of life, considering not only macroscopic, multicellular organisms but also microscopic extremophiles that thrive in extreme environments on Earth. The insights gained from studying extremophiles can offer valuable clues about the potential for life to exist in seemingly inhospitable environments elsewhere in the universe.

Our exploration will also encompass the ongoing search for exoplanets and the refinement of techniques for detecting biosignatures. The Kepler mission and its successors have revolutionized our understanding of planetary systems, revealing the remarkable prevalence of exoplanets. However, detecting biosignatures, which are subtle indicators of life within an exoplanet's atmosphere, requires extremely sensitive instrumentation and sophisticated data analysis techniques. The book will delve into the latest advancements in this area, discussing the challenges and potential breakthroughs in our ability to remotely detect evidence of life on other planets.

Furthermore, a significant portion of the book will be dedicated to critically evaluating reported UFO encounters. While many such accounts lack scientific credibility, some cases warrant careful examination, particularly those accompanied by compelling physical evidence or eyewitness testimonies. The key here is to adopt a rigorous, scientific approach, separating credible accounts from unsubstantiated claims through a meticulous analysis of evidence, taking into account potential misinterpretations, and acknowledging the limitations of observational data. Even anecdotal evidence, if meticulously documented and analyzed, can offer valuable clues and inform future research. The objective is not to confirm or deny the existence of extraterrestrial visitors, but to identify patterns, potential anomalies, and avenues for further scientific investigation. We will critically examine those accounts in light of known scientific phenomena, carefully considering alternative explanations before entertaining the possibility of extraterrestrial origins.

The book consciously avoids definitive statements about the existence of extraterrestrial life. Instead, its primary goal is to illuminate the scientific possibilities and challenges, presenting a reasoned argument based on the current state of scientific knowledge. This approach allows for a nuanced exploration of the topic, acknowledging the uncertainties and complexities involved while maintaining a focus on evidence-based reasoning. The conclusions drawn will be grounded in scientific evidence, acknowledging the gaps in our knowledge and highlighting areas requiring further research. This means that we will not shy away from exploring even the most speculative ideas if those ideas are grounded in scientific principles and testable hypotheses.

Central to this approach is the principle of scientific skepticism. While embracing the possibility of extraterrestrial life, we will maintain a healthy dose of skepticism, demanding robust evidence before drawing conclusions. This is not to say that skepticism is synonymous with dismissal; rather, it serves as a tool for critical evaluation, ensuring that our reasoning is sound and our conclusions are well-supported. Science thrives on skepticism, and it is this critical, questioning attitude that will guide our exploration of the evidence. We must strive for objectivity, avoiding confirmation bias and critically examining our own assumptions.

Finally, the book aims to stimulate broader discussion and engagement on the topic of extraterrestrial life. The renewed interest from governments worldwide and the increasing availability of data from space missions have created a unique opportunity for interdisciplinary collaboration and open dialogue among scientists, policymakers, and the public. Understanding the implications of discovering extraterrestrial life is crucial. It would undoubtedly reshape our understanding of our place in the universe and has profound implications for our future as a species. We will explore the ethical, societal, and philosophical ramifications of such a discovery. By engaging the reader in this multi-faceted examination, the book hopes to foster a deeper appreciation of the scientific endeavors involved and to encourage continued investigation and open discussion surrounding the possibility of life beyond Earth.

This interdisciplinary approach necessitates a commitment to collaboration. The exploration of extraterrestrial life transcends the boundaries of individual disciplines. We must leverage the expertise of astronomers, astrophysicists, biologists, geologists, chemists, and social scientists to comprehensively address the scientific, ethical, and societal dimensions of this topic. This book represents one attempt to bridge these diverse perspectives, drawing on the collective knowledge of these fields to present a comprehensive and well-rounded exploration of the possibility of extraterrestrial life.

The journey towards understanding the existence of extraterrestrial life is not merely a scientific endeavor but a profoundly human one. It reflects humanity's inherent curiosity, our insatiable quest for knowledge, and our fundamental desire to understand our place in the cosmos. The search itself, irrespective of its ultimate outcome, represents an inspiring testament to human ingenuity and our unwavering pursuit of answers to some of the most fundamental questions that humanity has ever asked. The scientific method, with its inherent emphasis on critical thinking, rigorous analysis, and a commitment to evidence-based reasoning, remains our most effective tool in navigating the vast unknown and uncovering the secrets of the universe. The journey promises to be both intellectually stimulating and deeply rewarding, regardless of whether we find definitive proof of life beyond Earth. The very process of searching has the potential to transform our understanding of ourselves and our place within the grand cosmic narrative.

| 2 |

Life's Extremes on Earth

The remarkable diversity of life on Earth extends far beyond the comfortable confines of temperate zones and lush ecosystems. A hidden world thrives in environments that would be instantly lethal to most organisms – a world populated by extremophiles. These remarkable creatures, defying our conventional understanding of life's limitations, push the boundaries of what we consider habitable, offering invaluable insights into the potential for life beyond Earth. Their existence suggests that life's adaptability might be far greater than we previously imagined, capable of flourishing under conditions we once deemed impossible.

Consider the boiling vents of deep-sea hydrothermal systems, where superheated water, rich in dissolved minerals, erupts from the Earth's crust. Here, in the crushing pressure and absolute darkness of the abyssal plain, thrive chemosynthetic organisms – bacteria and archaea that don't rely on sunlight for energy. Instead, they harness the chemical energy released from hydrothermal vents, oxidizing hydrogen sulfide and other chemicals to produce the energy necessary for life. These organisms form the base of unique ecosystems, supporting a complex food web that includes tube worms, giant clams, and other specialized creatures adapted to this extreme environment. The sheer pressure at these depths is immense, far exceeding anything experienced on the surface, yet these organisms not only survive but flourish. Their survival strategies, involving specialized proteins and cellular structures resistant to high pressure and temperature, provide a crucial template for understanding how life might adapt to similar conditions on other planets.

Moving from the crushing depths to the scorching heights, we encounter thermophiles, organisms that thrive in extremely high temperatures. These organisms, found in hot springs, geysers, and volcanic areas, possess enzymes and other proteins that remain stable and functional at temperatures that would denature proteins in most other organisms. Examples include members of the *Thermus* genus, bacteria that have been found in hot springs with temperatures exceeding 80°C. Their heat-stable DNA polymerases are now a staple in molecular biology laboratories worldwide, used in the polymerase chain reaction (PCR) technique, a fundamental tool in genetic research. The ability of these organisms to survive and reproduce at such high temperatures demonstrates the inherent robustness of life's underlying biochemistry and its potential to persist in environments that were once considered sterile.

Extremophiles are not limited to high temperatures; they also thrive in environments characterized by extreme cold, high salinity, and intense radiation. Psychrophiles, for instance, are organisms that thrive in extremely low temperatures, often below 0°C. These hardy organisms, found in glaciers, polar ice caps, and deep-sea sediments, possess adaptations that allow them to maintain cell membrane fluidity and enzymatic activity at sub-zero temperatures. Their cellular machinery has evolved to function effectively in an environment where water is often frozen, and metabolic processes are slowed dramatically. These adaptations provide compelling evidence for the potential of life to persist in the frigid environments of icy moons, like Europa or Enceladus, where subsurface oceans are believed to exist.

Halophiles, on the other hand, thrive in environments with extremely high salt concentrations, such as salt lakes and hypersaline ponds. These organisms have evolved unique mechanisms to maintain osmotic balance, preventing their cells from drying out in the hyper-saline environment. Their cell walls and membranes are modified to resist the high salt concentrations, and their metabolic pathways are adapted to function under these extreme conditions. The study of halophiles has revealed fascinating insights into the diversity of metabolic strategies that allow life to flourish in environments far removed from what we consider typical.

Another remarkable group of extremophiles are radioresistant organisms, capable of surviving extremely high doses of ionizing radiation. Deinococcus radiodurans, a bacterium discovered in irradiated canned meat, holds the record for radiation resistance. This organism possesses sophisticated DNA repair mechanisms that allow it to rapidly restore its genome after exposure to ionizing radiation. This exceptional ability raises intriguing possibilities about life's potential to thrive in the harsh radiation environments of space, perhaps within the shielded subsurface layers of planets or moons exposed to intense solar or cosmic radiation.

The study of extremophiles has profound implications for our understanding of the limits of life and the potential for life beyond Earth. Their adaptations to extreme environments challenge our Earth-centric view of habitability, suggesting that life might be far more adaptable and resilient than previously thought. The discovery of extremophiles living in conditions previously considered impossible has expanded the definition of "habitable zone" beyond simply the presence of liquid water, incorporating factors such as temperature, pressure, salinity, and radiation levels. The diversity of environments where extremophiles thrive highlights the possibility that life on other planets could employ different biochemistries, metabolic strategies, and cellular structures that differ significantly from those found on Earth.

Furthermore, understanding the biochemical mechanisms that allow extremophiles to survive in extreme environments provides crucial insights into the potential for life's emergence and evolution in diverse environments.

Their specialized enzymes and proteins, resistant to high temperatures, pressures, or radiation, could be of immense technological value. Heat-stable enzymes from thermophiles are already widely used in various industrial processes, highlighting the potential for bioprospecting – the search for useful products from extremophiles – to unlock new technological advancements. Moreover, extremophiles offer a unique window into early Earth conditions, as their biochemical processes and evolutionary strategies may mirror those of early life forms.

The study of extremophiles is not merely an academic pursuit; it has significant implications for several fields. Astrobiology, the study of life beyond Earth, relies heavily on the insights gained from studying extremophiles. Understanding how life can thrive in extreme terrestrial environments informs our search for extraterrestrial life, broadening our search criteria and strategies. The characteristics of extremophiles, their adaptations, and their survival mechanisms provide valuable clues for identifying potential biosignatures—indicators of life—on other planets. Furthermore, the study of extremophiles has applications in various technological fields. The development of new materials and processes inspired by the unique adaptations of these organisms could lead to groundbreaking technological advancements in fields such as biotechnology, bioremediation, and material science.

In summary, extremophiles represent a captivating realm of life on Earth, pushing the boundaries of our understanding of life's adaptability and resilience. Their presence in extreme environments serves as a powerful testament to the tenacity of life, expanding the possibilities of what we consider habitable, and significantly informing our ongoing quest to determine if we are alone in the universe. Their unique biochemical strategies, evolutionary adaptations, and technological potential continue to inspire wonder and fuel scientific investigation, promising further breakthroughs in our understanding of life's origins, its potential diversity, and its incredible capacity to adapt and thrive in the most unexpected of environments. The study of extremophiles is an ongoing journey of scientific discovery, a fascinating exploration into the extremes of life on Earth, and an invaluable tool in our search for life beyond.

The discovery and study of extremophiles have profoundly reshaped our understanding of the limits of life and, consequently, the potential for life beyond Earth. For decades, the search for extraterrestrial life was largely confined to the search for environments resembling Earth's – specifically, those with liquid water, moderate temperatures, and a protective atmosphere. The existence of extremophiles, however, shatters this anthropocentric view, demonstrating that life can thrive in environments previously deemed uninhabitable. This has broadened the scope of the search for extraterrestrial life significantly, extending the definition of a "habitable zone" far beyond its previously understood parameters.

The implications for astrobiology are immense. The discovery of extremophiles living in environments characterized by extreme temperatures, pressures, salinities, radiation levels, and pH values forces us to reconsider what constitutes a suitable environment for life. If life can flourish in the boiling vents of hydrothermal systems, the frigid depths of glacial ice, or the highly saline waters of salt lakes on Earth, it stands to reason that similar, or even more extreme, environments on other planets and moons could also harbor life.

This expanded definition of habitability opens up a wealth of possibilities in the search for extraterrestrial life. Planets and moons previously dismissed as uninhabitable based on their surface conditions, such as Europa or Enceladus with their subsurface oceans, now become compelling targets for exploration. The detection of hydrothermal vents on other celestial bodies, analogous to those found on Earth, would immediately raise the possibility of chemosynthetic ecosystems thriving beneath the icy surfaces. Similarly, the presence of subsurface liquid water, even if it is highly saline or exists under extreme pressure, becomes a potentially habitable environment, given the existence of halophiles and piezophiles (pressure-loving organisms) on Earth.

The study of extremophiles also provides crucial insights into the potential biosignatures of extraterrestrial life. Biosignatures are indicators of life, and the characteristics of extremophiles help us refine our search for these signals. For example, the unique metabolic processes of extremophiles may produce specific chemical byproducts that could be detected remotely through spectroscopic analysis of planetary atmospheres or subsurface materials. The detection of unusual isotopic ratios, for instance, could indicate the presence of life utilizing unique metabolic pathways, such as those found in methanogenic archaea living in anoxic environments. Similarly, the identification of specific organic molecules, or complex structures that exhibit chiral asymmetry (the presence of mirror-image molecules), could serve as strong indicators of past or present life.

The resilience of extremophiles to extreme conditions also offers valuable lessons for understanding the potential for life to survive extreme events, such as asteroid impacts or stellar flares. The radiation resistance of organisms like *Deinococcus radiodurans*, for instance, suggests that life might persist even in environments exposed to high levels of ionizing radiation. This has crucial implications for the potential for subsurface life on planets or moons that receive high levels of stellar or cosmic radiation. Their ability to repair damaged DNA quickly and efficiently makes them fascinating subjects of study, offering possible avenues for developing new radiation protection techniques, both for human space exploration and for protecting terrestrial life from radiation-related hazards.

Beyond the implications for astrobiology, extremophiles hold immense potential for technological innovation. Their unique adaptations have already led to significant advances in various fields. For instance, the heat-stable enzymes of thermophiles are widely used in various industrial applications, such as PCR technology and biofuel production. The potential for bioprospecting – the search for commercially valuable products from extremophiles – is enormous, with applications spanning biotechnology, bioremediation, and material science. Understanding the mechanisms by which extremophiles survive extreme conditions could lead to the development of new materials with enhanced properties, such as high-temperature lubricants or radiation-resistant polymers. This could revolutionize various industries, from aerospace to energy production.

Moreover, the study of extremophiles provides a crucial perspective on the early evolution of life on Earth. The harsh conditions prevalent on early Earth, characterized by high temperatures, intense volcanic activity, and a reducing atmosphere, likely favored the emergence of extremophilic life forms. Studying the metabolic pathways and genetic mechanisms of extant extremophiles provides insights into the potential characteristics of early life, helping us to understand how life arose from simple molecules and evolved to adapt to ever-changing environmental conditions. Understanding the evolution of extremophiles can also offer insight into the potential for life to evolve on other planets under diverse and challenging conditions. Their capacity to adapt to extreme conditions indicates that life may be far more versatile and adaptable than previously imagined.

Furthermore, the exploration of extremophiles expands our understanding of the fundamental principles of biology. Their unique adaptations challenge the conventional dogma surrounding the requirements for life, particularly the reliance on liquid water and specific temperature ranges. The discovery of organisms that thrive in the absence of oxygen, or under extremely high or low pH conditions, expands our definition of what constitutes a viable biosphere. This enhanced understanding is critical to informing the development of new hypotheses about life's origins and evolution, both on Earth and elsewhere in the universe. The inherent robustness and adaptability exhibited by extremophiles suggest that life, once initiated, may possess an extraordinary capacity for survival and diversification, even in the face of environmental catastrophes.

The implications of extremophile biology extend far beyond a simple expansion of the habitable zone. They represent a paradigm shift in our thinking about the nature of life itself. It's no longer a question of merely finding a planet with liquid water, but of identifying environments that harbor the chemical and energetic resources required for life, regardless of the specific environmental conditions. This opens the door to a vastly expanded search for life, beyond the conventional limitations of our terrestrial perspective. The discovery of extremophiles is not just a scientific achievement, but a philosophical one, forcing us to reconsider our place within the vast cosmos and the potential for life to exist in forms and environments far beyond our current understanding. The journey of discovery continues, promising further breakthroughs and a deepening appreciation for the remarkable tenacity and adaptability of life in the universe. The exploration of extremophiles is an ongoing, dynamic field, constantly pushing the boundaries of our knowledge and inspiring new avenues of research into the fascinating and potentially ubiquitous nature of life beyond Earth.

Analyzing the DNA and biochemistry of extremophiles reveals a remarkable array of adaptations that allow these organisms to thrive in environments lethal to most known life forms. These adaptations, honed over millions of years of evolution, offer profound insights into the fundamental principles of life and expand our understanding of the potential for life beyond Earth. The study primarily involves analyzing their genomes, which provide a blueprint for their survival strategies, and studying their unique metabolic pathways, which reveal how they extract energy and build essential components in their extreme habitats.

One crucial aspect is the analysis of extremophile DNA. Unlike mesophiles, organisms thriving in moderate conditions, extremophiles often possess modified DNA structures and repair mechanisms to withstand the damaging effects of their environments. For instance, thermophiles, thriving in high temperatures, often have more stable DNA structures, incorporating high proportions of guanine and cytosine base pairs, which form stronger hydrogen bonds than adenine and thymine pairs. These stronger bonds help maintain the integrity of the DNA molecule at elevated temperatures. Furthermore, many thermophiles possess specialized enzymes like reverse gyrases that introduce positive supercoils into their DNA, further stabilizing it against the heat-induced denaturation.

Another notable example is the extreme resistance of certain extremophiles to radiation. Organisms like *Deinococcus radiodurans*, a polyextremophile that can withstand extreme radiation, desiccation, and oxidative stress, possess multiple copies of their genome and highly efficient DNA repair mechanisms. When exposed to radiation, their genome fragments extensively, but remarkably, they efficiently recombine these fragments to restore the original genome, demonstrating an exceptional capacity for damage repair. Studying the proteins involved in this repair process—particularly those involved in homologous recombination and non-homologous end joining—holds enormous implications for biotechnology, including potential applications in radiation protection technologies.

Beyond DNA structure, the analysis of extremophile genomes reveals unique genetic features reflecting adaptation to specific environmental stressors. Halophiles, organisms thriving in high-salt environments, often have modified proteins and membranes to maintain osmotic balance. Their proteins often contain a high proportion of acidic amino acids, preventing aggregation and maintaining function under high salinity. Their cell membranes are also unique, frequently utilizing specialized lipids that are more resistant to disruption by high salt concentrations. Analyzing the genes encoding these specialized proteins and lipids unveils the genetic basis of this remarkable adaptation.

Similarly, psychrophiles, thriving in extremely cold environments, have evolved mechanisms to maintain enzyme activity at low temperatures. Their enzymes often have more flexible structures, allowing for catalysis at low temperatures, and their membranes contain unsaturated fatty acids, which maintain fluidity at freezing temperatures. The genes encoding these cold-adapted enzymes are of immense interest to various industries, including biofuel production and low-temperature industrial processes.

Analyzing the biochemistry of extremophiles is equally revealing. These organisms often employ unique metabolic strategies to exploit available energy sources in their environments. Chemoautotrophs, for example, are prevalent in deep-sea hydrothermal vents, where they utilize inorganic chemicals like hydrogen sulfide as an energy source to fix carbon dioxide into organic molecules. Studying the enzymes and pathways involved in this chemosynthesis provides a fascinating glimpse into the diversity of metabolic strategies and the potential for life to thrive in environments devoid of sunlight. This is particularly crucial for understanding the potential for life on planets or moons with limited or no access to solar energy.

Methanogens, a type of archaea, are another significant group of extremophiles that employ unique metabolic strategies. They thrive in anaerobic environments, producing methane as a byproduct of their metabolism. Understanding the mechanisms by which they produce and utilize methane is crucial for astrobiological research, as methane has been detected on several planets and moons, and its presence can potentially indicate the presence of methanogenic life. The analysis of their metabolic pathways is also important for studying carbon cycling on Earth and other planetary bodies.

Acidophiles and alkaliphiles, organisms that thrive in extremely acidic or alkaline environments, respectively, present unique challenges in maintaining intracellular pH. These organisms employ various strategies to regulate their internal pH, including sophisticated proton pumps and buffering systems. Understanding these mechanisms provides insights into the limits of life's ability to regulate its internal environment. Analyzing the genes and proteins responsible for pH homeostasis could shed light on the fundamental processes underlying this crucial adaptation.

Beyond their unique adaptations, extremophiles also exhibit remarkable resilience and adaptability. Many extremophiles are polyextremophiles, capable of withstanding multiple stressors simultaneously, highlighting the interconnected nature of these adaptations. This remarkable resilience has profound implications for understanding the persistence and evolution of life in environments subject to extreme fluctuations. Studying their ability to adapt to changing environmental conditions could inform research into the development of resilient crops and other organisms capable of withstanding climate change impacts.

The study of extremophile DNA and biochemistry is not limited to their individual adaptations but extends to understanding their evolutionary history. Comparative genomics of various extremophile lineages reveals the evolutionary trajectories leading to their remarkable survival strategies. Understanding how these adaptations evolved provides insights into the fundamental evolutionary processes shaping life on Earth and may offer clues to potential evolutionary pathways on other planets. By comparing the genomes of extremophiles from various environments, researchers can determine the commonalities and differences in their evolutionary adaptations, shedding light on the constraints and opportunities presented by different extreme environments.

The development of new high-throughput sequencing technologies has significantly advanced the study of extremophile genomes. These technologies allow researchers to sequence thousands of extremophile genomes simultaneously, dramatically increasing the speed and efficiency of genomic studies. Sophisticated bioinformatics tools are crucial in analyzing the vast amounts of data generated by these sequencing efforts, allowing researchers to identify genes and pathways crucial for extremophile survival.

The development of advanced imaging techniques, such as cryo-electron microscopy, is also advancing our understanding of extremophile biochemistry. These techniques allow researchers to visualize the three-dimensional structures of proteins and other macromolecules at atomic resolution, revealing insights into how these molecules function in extreme environments. Combining genomics and proteomics data provides a comprehensive view of the intricate network of interacting genes and proteins responsible for extremophile adaptation. Advanced metabolomics techniques, allowing the identification and quantification of a wide range of metabolites, are also crucial in understanding extremophile metabolic pathways. Combining these approaches paints a more comprehensive picture than previously possible.

In conclusion, the analysis of extremophile DNA and biochemistry has revolutionized our understanding of the limits of life and the potential for life beyond Earth. The unique adaptations of these organisms, revealed through genomic and biochemical analyses, challenge the traditional notions of what constitutes a habitable environment. The information gleaned from studying these remarkable organisms is not only fundamental to astrobiology but also holds immense potential for technological advancements across diverse fields, from biotechnology to material science, significantly expanding our knowledge and capabilities. The ongoing research in this vibrant field continues to yield remarkable discoveries, pushing the boundaries of our understanding of the extraordinary diversity and resilience of life on Earth and its implications for life beyond our planet.

Extrapolating the lessons learned from Earth's extremophiles to other planetary bodies requires a careful consideration of the environmental parameters present on those bodies. While we haven't yet discovered extraterrestrial life, the remarkable adaptability of extremophiles significantly expands the range of environments we consider potentially habitable. This involves comparing the physical and chemical conditions of extreme environments on Earth with those found on other planets and moons within our solar system and beyond.

One compelling parallel is the comparison between hydrothermal vents on Earth and those potentially present on other celestial bodies, such as Europa (Jupiter's moon) and Enceladus (Saturn's moon). Hydrothermal vents on Earth support thriving ecosystems completely independent of sunlight, relying instead on chemosynthesis. The discovery of subsurface oceans on Europa and Enceladus, coupled with evidence suggesting hydrothermal activity on these moons, fuels speculation about the existence of similar chemosynthetic ecosystems. The presence of water, along with chemical energy sources like hydrothermal vents, could provide the necessary ingredients for life to flourish even in the absence of sunlight. The potential for life in these subsurface oceans is particularly exciting because it suggests that life could exist in places previously considered uninhabitable. Further exploration and investigation, including missions specifically designed to analyze the composition of these subsurface oceans, are crucial to determining the habitability of these icy moons.

The existence of subsurface liquid water is not limited to Europa and Enceladus. Mars also shows evidence of past liquid water on its surface, and the possibility of subsurface liquid water persists. Though the Martian surface is presently inhospitable due to its extremely cold and dry conditions, and the intense radiation, extremophiles that can tolerate desiccation, radiation, and low temperatures could potentially survive in subsurface environments. Studies of extremophiles such as *Deinococcus radiodurans*, with its extraordinary radiation resistance, suggest that life could persist in the Martian subsurface even if exposed to high levels of radiation. The search for signs of past or present life on Mars, therefore, focuses heavily on exploring subsurface habitats. Missions like the Mars 2020 Perseverance rover are specifically designed to search for biosignatures in ancient riverbeds and other locations where liquid water might have existed in the past.

Similarly, Titan, Saturn's largest moon, presents a unique case. While it lacks liquid water on its surface, it possesses vast lakes and seas of liquid methane and ethane. The possibility of life utilizing methane-based biochemistry on Titan is a captivating astrobiological question. Though this would represent a fundamental divergence from known terrestrial life, the existence of methanogenic extremophiles on Earth shows that life can utilize diverse carbon sources. Studying the metabolic pathways of methanogens could shed light on the potential for life to thrive in environments drastically different from our own. Future missions to Titan, such as the Dragonfly rotorcraft lander, are designed to investigate the organic chemistry and potentially search for signs of life in Titan's hydrocarbon-rich environment.

Beyond our solar system, the discovery of exoplanets—planets orbiting stars other than our Sun—further expands the possibilities for extraterrestrial life. The detection of exoplanets within their star's habitable zone, where liquid water could potentially exist on the planet's surface, has become increasingly common. However, defining the habitable zone is complex and depends on several factors, including the star's luminosity and the planet's atmospheric composition. Extremophiles on Earth demonstrate that the definition of "habitable" needs to be expanded beyond the traditional notion of a planet with liquid water at a moderate temperature. The presence of extremophiles in highly saline, acidic, or alkaline environments, as well as in extremely cold or hot environments, demonstrates the remarkable adaptability of life and suggests that life could exist in a far wider range of environments than previously considered.

The study of extremophiles also has significant implications for the search for extraterrestrial biosignatures. Traditional methods of searching for extraterrestrial life focus on detecting molecules associated with known terrestrial life, such as oxygen or specific organic molecules. However, the diversity of metabolic pathways employed by extremophiles suggests that alien life may utilize different metabolic strategies, leading to the production of unique biosignatures. Understanding the potential range of biosignatures produced by different types of extremophiles is therefore crucial for designing more effective strategies for detecting extraterrestrial life. This includes expanding the search beyond the traditional markers of life and developing new methods for detecting potentially alien biomarkers.

Moreover, the extremophile examples highlight the importance of considering the interconnected nature of environmental factors in determining habitability. Many extremophiles are polyextremophiles, able to tolerate multiple stressors simultaneously. This interconnectedness implies that the survival of life in an extreme environment may depend on the synergistic interactions between several environmental factors. For instance, the presence of a subsurface ocean on an icy moon might not be sufficient for life to exist if the ocean lacks sufficient chemical energy sources or if the overlying ice layer is too thick to allow for sufficient exchange of materials with the surrounding environment. Analyzing these interactions is important for developing more comprehensive models of planetary habitability.

Furthermore, extremophiles' resilience to environmental change has implications for understanding the long-term evolution of life on other planets. Many extremophiles are remarkably adaptable, able to survive and even thrive in environments that undergo substantial change over time. This resilience might be crucial for life to survive through periods of climate change or other environmental fluctuations. Studying how extremophiles adapt to such changes could inform the development of models that predict the likelihood of life's persistence on planets that experience significant environmental variability.

Finally, the study of extremophiles directly informs the search for life beyond Earth by providing testable hypotheses about the potential for life in extreme environments. The unique adaptations and survival strategies of these organisms provide a framework for developing new experiments and technologies to search for life in environments that may initially seem uninhabitable. For example, the development of instruments capable of detecting unique biosignatures generated by extremophile metabolism could significantly enhance the search for extraterrestrial life. In the coming decades, missions to other planetary bodies, driven by the lessons learned from studying extremophiles, will continue to push the boundaries of our understanding of the potential for life beyond Earth. The more we understand the limits of life here on Earth, the more effectively we can search for the potential existence of life elsewhere.

Our understanding of habitability has traditionally been heavily influenced by Earth-centric biases. The "habitable zone," often depicted as a narrow band around a star where liquid water can exist on a planet's surface, is a prime example. This concept, while useful as a starting point, is ultimately a simplification. It assumes that liquid water at temperatures conducive to known terrestrial life is the absolute prerequisite for life's existence. However, the remarkable diversity of extremophiles on Earth fundamentally challenges this narrow definition. Their survival in environments previously considered uninhabitable forces us to reconsider the fundamental parameters defining a habitable world.

The discovery of extremophiles thriving in extreme conditions—intense heat, extreme cold, high acidity, intense radiation, or extreme salinity—suggests that life's capacity for adaptation far surpasses what we once believed. These organisms have evolved ingenious strategies to cope with environmental stressors, utilizing unique biochemical pathways and cellular structures to survive and even thrive in seemingly impossible environments. For example, the discovery of hyperthermophiles in hydrothermal vents, flourishing at temperatures exceeding the boiling point of water, demonstrates that life can not only tolerate but also prosper in conditions that would instantly kill most known terrestrial organisms. Similarly, psychrophiles, adapted to sub-zero temperatures, thrive in polar regions and deep-sea environments. Halophiles, thriving in hypersaline environments like the Great Salt Lake, demonstrate life's ability to adapt to extreme salinity. These are just a few examples of the incredible resilience and adaptability displayed by life on Earth.

The implications of extremophile discoveries for astrobiology are profound. The existence of such organisms suggests that the search for extraterrestrial life should not be confined to environments that closely resemble Earth's surface conditions. Instead, we must broaden our search to encompass a wider range of environments, including those previously deemed inhospitable. This includes considering planets and moons orbiting stars other than our Sun—exoplanets—which may possess radically different atmospheric compositions, gravitational forces, and geological histories than Earth.

Consider the potential for life on planets orbiting red dwarf stars. Red dwarfs are the most common type of star in the Milky Way galaxy, and their smaller size and lower luminosity mean that their habitable zones are much closer to the star than Earth's. Planets within these habitable zones would experience tidal locking, with one side permanently facing the star and the other side perpetually in darkness. This extreme temperature gradient might seem inhospitable, but it doesn't preclude the possibility of life. Extremophiles on Earth have demonstrated the ability to cope with dramatic temperature variations. Life on a tidally locked planet might evolve to thrive in the twilight zone between the perpetually light and dark sides, or it could adapt to the extreme conditions on one side or the other.

Further broadening our horizons, we must also consider the possibility of life in subsurface environments. The discovery of subsurface oceans on several icy moons within our solar system, such as Europa and Enceladus, has revolutionized our understanding of the potential for extraterrestrial life. These subsurface oceans, shielded from the harsh radiation of space, could potentially harbor life, similar to the hydrothermal vent ecosystems on Earth. The energy source for these ecosystems could be chemosynthesis, based on chemical reactions rather than sunlight. The presence of hydrothermal vents, fueled by the moon's internal heat, could provide the necessary chemical energy for life to thrive in these dark, subterranean environments.

Expanding the definition of habitability also requires us to reconsider the role of water. While liquid water is considered essential for life as we know it, the possibility of life based on different solvents cannot be ruled out. Titan, Saturn's largest moon, possesses vast lakes and seas of liquid methane and ethane. The potential for life utilizing methane-based biochemistry on Titan is a particularly intriguing question. While this would represent a significant departure from terrestrial life, the discovery of methanogenic extremophiles on Earth provides a compelling precedent. These organisms utilize methane as an energy source, suggesting that life can evolve to utilize diverse carbon sources.

Furthermore, the limitations of our current understanding of biochemistry must be acknowledged. Our current understanding of life is largely based on terrestrial organisms, and we may be severely underestimating the diversity of life's potential forms. The search for extraterrestrial life should not be limited to searching for organisms that resemble terrestrial life. We need to develop new methods to detect and identify life based on different biochemical principles.

This necessitates a reassessment of biosignatures, the chemical or physical indicators of past or present life. Traditional approaches to searching for extraterrestrial life often focus on detecting biosignatures that are characteristic of terrestrial life, such as oxygen or specific organic molecules. However, given the diversity of extremophiles and the possibility of life based on different biochemistries, we need to broaden our search to encompass a far wider range of potential biosignatures. This requires interdisciplinary collaborations between astrobiologists, chemists, geologists, and other scientists to develop new detection methods and refine our understanding of what constitutes a biosignature.

The search for extraterrestrial life is not simply a search for a "second Earth." It's a quest to understand the full extent of life's potential across the cosmos. The remarkable adaptations of extremophiles on Earth provide valuable insights into the potential for life to exist in environments far beyond our current conceptions of habitability. By expanding our definition of habitability, and developing new techniques for detecting diverse forms of life, we can significantly enhance our chances of discovering life beyond Earth and gain a deeper appreciation of life's boundless creativity and resilience. The exploration of extreme environments on Earth is not merely a scientific pursuit; it is a vital step in answering one of humanity's most fundamental questions: are we alone in the universe? The answer, as the diversity of extremophiles convincingly demonstrates, may lie in environments far more diverse and unexpected than we once imagined. The universe, it appears, may hold far more surprises than we are currently equipped to comprehend. The expansion of our definition of habitability is not just an academic exercise; it's a necessary step in broadening our search and our understanding of the vast and potentially life-filled cosmos.

| 3 |

The Chemistry of Life

Life, as we know it, is fundamentally a chemical phenomenon. Understanding the building blocks of life, therefore, requires a deep dive into the world of chemistry, exploring the elements, molecules, and reactions that make life possible. While the search for extraterrestrial life often focuses on grand narratives of habitable zones and alien civilizations, the very foundation of life's possibility rests on the intricate dance of atoms and molecules. Let's begin by examining the elements most crucial to life as we understand it.

The periodic table, a seemingly abstract arrangement of elements, becomes a vital roadmap when exploring the chemistry of life. While the table boasts over 100 elements, only a handful are truly essential for the construction of living organisms. Carbon, hydrogen, oxygen, nitrogen, phosphorus, and sulfur—often abbreviated as CHNOPS—are the workhorses of biological chemistry. Carbon, with its remarkable ability to form four strong covalent bonds, stands out as the central element of life. This tetravalence allows carbon to create a vast array of complex molecules, forming the backbone of carbohydrates, lipids, proteins, and nucleic acids—the four major classes of biological macromolecules. These macromolecules are the building blocks of cells, tissues, and organs, performing a multitude of functions essential for life.

Hydrogen, the smallest and most abundant element in the universe, is ubiquitous in biological molecules. It's a key component of water, the universal solvent, and is found within countless organic molecules, contributing to their structure and reactivity. Oxygen, crucial for aerobic respiration, plays a vital role in energy production in most terrestrial organisms. It's also a key component of water and many organic molecules. Nitrogen, integral to amino acids (the building blocks of proteins) and nucleic acids (DNA and RNA), is essential for genetic information storage and protein synthesis. Phosphorus, primarily found in the backbone of DNA and RNA, and in ATP (adenosine triphosphate), the energy currency of cells, is vital for heredity and energy transfer. Lastly, sulfur, a crucial component of many proteins, plays a critical role in maintaining their three-dimensional structure and function. While other elements like iron, magnesium, calcium, and potassium are also essential for life, albeit in smaller amounts, CHNOPS form the fundamental chemical architecture of living organisms.

Beyond these essential elements, the role of water cannot be overstated. Water's unique properties are intimately linked to its molecular structure. The polar nature of the water molecule, with its slightly positive and negative ends, allows it to form hydrogen bonds with other water molecules and with other polar molecules. This leads to water's high surface tension, its ability to act as a solvent for a wide range of substances, its high specific heat capacity (meaning it resists temperature changes), and its high heat of vaporization (meaning it requires significant energy to evaporate). These properties are critical for maintaining stable internal environments within living organisms and for various biological processes. The solvent properties of water are particularly important, as they allow for the transport of nutrients and waste products within cells and throughout organisms.

The remarkable versatility of carbon extends beyond its presence in the CHNOPS elements. Carbon atoms can bond together to form long chains, branched structures, and rings, creating an incredible diversity of organic molecules. These molecules form the basis of life's complexity, from simple sugars to complex proteins and nucleic acids. Carbohydrates, providing energy and structural support, are composed of carbon, hydrogen, and oxygen atoms. Lipids, including fats and oils, serve as energy storage molecules, form cell membranes, and act as hormones. They are largely composed of carbon and hydrogen, with relatively fewer oxygen atoms. Proteins, built from chains of amino acids, are responsible for a wide array of functions, including catalysis, transport, structural support, and immune defense. Their intricate three-dimensional structures, dictated by the sequence of amino acids, are crucial for their function. Lastly, nucleic acids, DNA and RNA, store and transmit genetic information, governing the synthesis of proteins and the inheritance of traits. These molecules contain carbon, hydrogen, oxygen, nitrogen, and phosphorus.

The intricacies of biological molecules extend beyond their composition to encompass their remarkable three-dimensional structures. Proteins, for instance, fold into precise shapes based on interactions between their amino acids, including hydrogen bonds, hydrophobic interactions, and disulfide bridges. This precise folding is critical for their function. Similarly, the double helix structure of DNA, stabilized by hydrogen bonds between complementary base pairs, is essential for its role in storing and transmitting genetic information. The complex interactions between these molecules and their specific spatial arrangements are crucial to the organization and functionality of cells and organisms.

Understanding the fundamental chemistry of life, therefore, necessitates a broader perspective than simply listing the elements involved. It encompasses the intricate ways in which these elements combine, the properties of the resulting molecules, and the complex interactions between them. The precise three-dimensional structures of biological molecules are a testament to life's extraordinary sophistication, arising from the fundamental laws of physics and chemistry. This precise molecular orchestration highlights the delicate balance of forces, and chemical reactions that underpin the incredible complexity of life.

Considering the vastness of the universe, the possibility of life existing beyond Earth based on fundamentally different chemical principles is a compelling prospect. While CHNOPS and water may be vital to terrestrial life, it's crucial to consider the potential for alternative biochemistries. The discovery of extremophiles on Earth, flourishing in seemingly impossible environments, has already challenged our preconceived notions of habitability. Perhaps life on other planets might utilize different solvents, different elements, or different molecular structures entirely. It's imperative that our search for extraterrestrial life be informed by a deep understanding of the fundamental chemical principles governing life's emergence and diversity, acknowledging the possibility of variations far beyond our current understanding. The pursuit of understanding the building blocks of life is not merely an academic exercise; it is a crucial step in broadening our search for life beyond Earth and expanding our understanding of life's potential throughout the cosmos. The more we learn about the chemistry of life on Earth, the more effectively we can interpret data gathered from space missions exploring other celestial bodies. The search for extraterrestrial life requires a shift away from anthropocentric biases, allowing us to entertain the possibility of life forms vastly different from anything we've encountered on Earth.

This exploration into the chemical foundations of life underscores the crucial link between chemistry, biology, and astrobiology. As we broaden our understanding of the fundamental principles governing life's emergence and diversity on Earth, we simultaneously expand the scope of our search for life beyond our planet. The quest for extraterrestrial life is ultimately a quest for a deeper understanding of life itself, challenging our preconceived notions and pushing the boundaries of scientific inquiry. The universe may hold life forms that defy our current understanding, and the more we learn about the fundamental chemistry of life, the better equipped we will be to identify and understand them. The chemical exploration of life is a journey of discovery, leading us to a deeper appreciation of the intricate processes that underpin existence and the vast potential for life in the universe.

Our current understanding of life is heavily biased by our terrestrial experience. We know life on Earth thrives on carbon-based chemistry, utilizing water as a solvent. This CHNOPS-water paradigm, while undeniably successful on our planet, might not be universally applicable across the cosmos. The vastness and diversity of the universe suggest that life, if it exists elsewhere, may have evolved utilizing completely different biochemical pathways and elemental building blocks. This possibility opens up a fascinating realm of speculation and research into alternative biochemistries.

One of the most frequently discussed alternatives to carbon-based life is silicon-based life. Silicon, located directly beneath carbon on the periodic table, shares some similarities in its chemical behavior. Like carbon, silicon can form four covalent bonds. However, there are crucial differences that impact its suitability as a basis for life.

Silicon-silicon bonds are significantly weaker than carbon-carbon bonds, making the construction of long, stable chains—essential for complex biological molecules like proteins and nucleic acids—far more challenging. Moreover, silicon-oxygen bonds are exceptionally strong, leading to the formation of stable silicates, which are generally insoluble in water. This insolubility poses a significant hurdle for the efficient transport and utilization of nutrients essential for life. While silicon-based life remains a captivating hypothetical possibility, the challenges presented by the chemical properties of silicon suggest that it's unlikely to underpin life as complex as what we see on Earth.

Beyond silicon, other elements have been proposed as potential building blocks for extraterrestrial life. Some researchers have explored the possibility of life based on phosphorus or arsenic. While phosphorus is essential for life as we know it, particularly in DNA and RNA, arsenic, which is chemically similar, could theoretically substitute for it in some biological molecules. However, arsenic's toxicity to most organisms raises significant questions about its suitability as a fundamental building block for life. The successful incorporation of arsenic into the biochemistry of a living organism would require a fundamentally different metabolic pathway, far removed from what we observe on Earth. The discovery of a bacterium, *GFAJ-1*, capable of incorporating arsenic into its DNA, briefly ignited excitement about arsenic-based life, but subsequent research largely debunked this notion, highlighting the intricacies and limitations of replacing fundamental elements.

The solvent used by life is another crucial aspect of biochemistry. Water, with its exceptional solvent properties and its role in numerous biological reactions, is indispensable for terrestrial life. However, other solvents might support life under different environmental conditions. Ammonia, for example, is a liquid at lower temperatures than water and could potentially serve as a solvent in extremely cold environments. While ammonia-based life presents many challenges, including the lower stability of many organic molecules in ammonia, it remains a tantalizing theoretical possibility for life on icy moons or planets. Methane, another potential solvent, could support life at extremely low temperatures, opening up the possibility of life existing on some of the icy moons of our outer solar system. The key here is that the solvent needs to be capable of dissolving and transporting essential molecules for metabolic processes, while also allowing for the maintenance of a stable internal environment.

The concept of alternative biochemistries also extends to the types of molecules that could form the basis of life. While proteins, carbohydrates, lipids, and nucleic acids are the workhorses of terrestrial biochemistry, other molecular structures could theoretically perform similar functions. For instance, some researchers have proposed the possibility of life utilizing different polymers, such as alternative types of nucleic acids or different types of lipid membranes. Such variations could dramatically alter the way life functions and appears, rendering it virtually unrecognizable compared to terrestrial life forms.

The search for extraterrestrial life often focuses on the search for biosignatures – indicators of life – that we understand based on our terrestrial experience. However, the very notion of a biosignature needs to be broadened to include potentially alien biochemistries. This necessitates a multifaceted approach, including the development of sophisticated instruments capable of detecting a wider range of chemical compounds and the reassessment of what constitutes "life" itself. The search for extraterrestrial life requires us to move beyond anthropocentric biases and to consider the possibility of life based on radically different biochemical processes and elemental compositions.

Furthermore, the environment plays a critical role in shaping the biochemistry of life. Extremophiles on Earth, organisms thriving in environments considered inhospitable for most life, demonstrate the remarkable adaptability and diversity of life. These organisms, found in deep-sea hydrothermal vents, highly acidic or alkaline environments, or under extreme temperatures and pressures, offer valuable insights into the potential for life in seemingly uninhabitable conditions. Studying these extremophiles allows us to expand our understanding of life's potential limits and the range of possible biochemistries beyond the confines of our own planet. Their existence challenges the narrow definition of "habitable zones" traditionally used in the search for extraterrestrial life and implies that environments previously considered too extreme for life may, in fact, harbor unique and unexpected forms.

The challenge of identifying alternative biochemistries is immense, requiring a shift in our thinking about the prerequisites for life. We need to develop new theoretical frameworks and experimental techniques to explore these possibilities. Computational models can aid in the investigation of different biochemical pathways, and advanced analytical instruments are crucial for detecting subtle yet significant chemical signatures of alternative life forms. The integration of various scientific disciplines, including chemistry, biology, physics, and computer science, will be crucial for understanding the vast possibilities of extraterrestrial life.

The discovery of extraterrestrial life, regardless of its biochemistry, would revolutionize our understanding of the universe and our place within it. It would not only confirm that life is not a unique phenomenon confined to Earth but also fundamentally alter our scientific paradigms and our philosophical perspectives. The implications would be far-reaching, impacting fields from biology and astronomy to sociology and philosophy. It would reshape our understanding of evolution, the origin of life, and the potential for life elsewhere in the cosmos. The pursuit of understanding alternative biochemistries is therefore not simply an academic exercise but a vital step towards a broader and more complete understanding of life itself, expanding our horizons and pushing the limits of scientific exploration. The possibility of diverse, unconventional life forms challenges us to question our assumptions and to embrace the immense unknown that awaits us in the vast expanse of the universe. The search for life beyond Earth is, at its core, a quest to better understand life's fundamental nature and its capacity for adaptation and innovation, both here and in places we have yet to explore. The more we learn about the possibilities of life on Earth, and the potential for different chemistries to support it, the more prepared we will be to discover and interpret evidence of life beyond our own planet. This ongoing exploration is a testament to humanity's relentless curiosity and our unwavering pursuit of knowledge.

The question of life's origin on Earth remains one of science's most enduring mysteries. While we have a detailed understanding of evolution and the diversification of life since its emergence, the precise mechanisms that led to the very first self-replicating molecules remain elusive. Several compelling hypotheses attempt to explain this pivotal event, each with its own strengths and weaknesses, its own set of supporting evidence, and its own collection of unanswered questions.

One dominant hypothesis is abiogenesis, the process by which life arises from non-living matter. This is not a single, unified theory but rather a collection of models proposing different pathways through which simple organic molecules could have spontaneously assembled into more complex structures, eventually leading to self-replicating systems capable of evolution. The early Earth offered a very different environment than what we experience today. The atmosphere was likely rich in reducing gases like methane, ammonia, and hydrogen, with little to no free oxygen. This environment, far removed from our current oxygen-rich atmosphere, might have been more conducive to the formation of organic molecules from inorganic precursors.

The Miller-Urey experiment, conducted in 1952, provided crucial early support for abiogenesis. Stanley Miller and Harold Urey simulated the conditions of early Earth in a laboratory setting, passing electrical discharges through a mixture of gases representing the early atmosphere. The experiment successfully produced several amino acids, the building blocks of proteins, demonstrating that organic molecules could form spontaneously under prebiotic conditions. While the precise composition of the early Earth's atmosphere is still debated, the Miller-Urey experiment demonstrated the plausibility of abiogenesis and sparked intense research into the formation of organic molecules under various prebiotic conditions. Subsequent experiments have replicated and expanded on this work, producing a wider array of organic molecules under different environmental conditions, further solidifying the hypothesis that life's building blocks could have formed naturally on early Earth.

However, the formation of simple organic molecules is only the first step in the complex process of abiogenesis. The next challenge lies in understanding how these molecules assembled into more complex structures, such as proteins, nucleic acids, and lipid membranes, essential for the formation of cells. One proposed mechanism is self-assembly, where simple molecules spontaneously aggregate into larger structures driven by non-covalent interactions such as hydrogen bonds and van der Waals forces. Hydrothermal vents, deep-sea volcanic systems spewing hot, mineral-rich fluids, are considered potential sites for this process. The chemical gradients and catalytic properties of these vents might have provided the necessary conditions for the formation of complex organic molecules and their self-assembly into protocells, precursor structures to the first living cells. The discovery of extremophiles—organisms thriving in extreme environments—in and around these vents underscores the possibility of such a scenario. These organisms highlight the versatility and adaptability of life and broaden our understanding of the environmental conditions where life can emerge.

Another crucial aspect of abiogenesis is the origin of self-replication. The ability to replicate genetic information is a defining characteristic of life. RNA, a simpler molecule than DNA, is thought to have played a crucial role in early life. The RNA world hypothesis proposes that RNA, not DNA, was the primary genetic material in early life, capable of both storing genetic information and catalyzing chemical reactions. This hypothesis is supported by the fact that some RNA molecules, called ribozymes, possess catalytic activity, suggesting that early life could have relied on RNA alone for both information storage and catalysis before the evolution of more complex DNA-based systems. However, the exact transition from an RNA world to a DNA-based world remains an area of active research. The complex interactions and precise steps involved in this transition are still being elucidated.

The hypothesis of panspermia offers a dramatically different perspective on the origin of life. Panspermia proposes that life originated elsewhere in the universe and was transported to Earth. This hypothesis suggests that life's building blocks or even primitive life forms could have arrived on Earth via meteorites, comets, or other celestial bodies. The discovery of organic molecules in meteorites supports this idea. These molecules, including amino acids and nucleobases, show that the building blocks of life can exist in extraterrestrial environments. While this doesn't directly prove panspermia, it strongly suggests that the raw materials for life are not unique to Earth and could have been delivered to our planet from elsewhere in the universe.

There are two main variations within the panspermia hypothesis: lithopanspermia and ballistic panspermia. Lithopanspermia proposes that life arrived on Earth embedded within rocks ejected from other planets or moons. These rocks could have been propelled into space by asteroid impacts and travelled through interstellar space before eventually landing on Earth. Ballistic panspermia, on the other hand, focuses on the transfer of life between planets within the same solar system. This mechanism suggests that life could have spread from one planet to another through the ejection and subsequent transfer of material through asteroid impacts or other celestial events.

However, panspermia merely shifts the question of life's origin to another location. It doesn't explain *how* life arose initially, but rather suggests a mechanism for its distribution across the cosmos. The challenges of survival during interstellar or even interplanetary travel, including exposure to radiation and extreme temperatures, pose significant hurdles for this hypothesis. Despite these challenges, panspermia remains a viable and fascinating hypothesis. The constant bombardment of Earth by meteorites, and the discovery of organic molecules within them, lend considerable credence to the possibility of extraterrestrial origins for life's building blocks, if not life itself.

Understanding the origin of life on Earth is a multidisciplinary endeavor, demanding expertise in chemistry, biology, geology, and astronomy. Sophisticated laboratory experiments, the analysis of ancient rocks and meteorites, and the study of extremophiles provide crucial pieces of this complex puzzle. Advanced computational modeling techniques are also increasingly essential for simulating prebiotic conditions and testing various hypotheses about the early steps in the formation of life. While no single theory fully explains the origin of life, a combination of abiogenesis and panspermia might provide the most comprehensive explanation. The possibility that life's building blocks originated elsewhere in the universe and then assembled into self-replicating systems on early Earth is a captivating and plausible scenario. The ongoing research into these hypotheses is crucial not only for understanding the origin of life on Earth but also for guiding our search for extraterrestrial life. The broader implications of discovering life elsewhere in the universe are profound and far-reaching, touching on our understanding of biology, cosmology, and even our place in the cosmos. The pursuit of understanding life's origins is a continuous journey of exploration and discovery, constantly refining our models and challenging our assumptions as new data emerges. The mystery of life's origin remains a compelling challenge, urging continued investigation and scientific innovation in this exciting field.

Applying the principles gleaned from studying the origin of life on Earth significantly enhances our ability to search for life elsewhere in the cosmos. The diverse array of life forms inhabiting our planet, from the extremophiles thriving in the most inhospitable environments to the complex ecosystems of rainforests and coral reefs, underscores the remarkable adaptability and resilience of life. This biodiversity suggests that life, once initiated, can diversify and flourish under a wide range of conditions, potentially far exceeding what we initially consider habitable. Therefore, the search for extraterrestrial life should not be limited to environments mirroring Earth's conditions, but instead should encompass a broader spectrum of potential habitats.

One key takeaway from studying terrestrial life is the surprising prevalence of extremophiles. These organisms, capable of surviving and even thriving in extreme environments—extreme temperatures, high pressures, high salinity, extreme pH levels, and intense radiation—expand our understanding of life's limits. The discovery of extremophiles living near hydrothermal vents deep in the ocean, within highly acidic or alkaline environments, and even within radioactive waste, challenges our anthropocentric view of habitability. It suggests that the definition of a "habitable zone" may need to be reassessed and broadened to encompass environments previously deemed uninhabitable. Applying this broadened perspective to exoplanet research opens up the possibility of finding life in previously overlooked locations within planetary systems.

For instance, the moons of the outer solar system, such as Europa (Jupiter's moon) and Enceladus (Saturn's moon), are now considered prime targets in the search for extraterrestrial life. These icy moons possess subsurface oceans, potentially containing liquid water, a crucial ingredient for life as we know it. The energy sources for these subsurface oceans may be tidal forces from their respective planets, generating heat through friction, or hydrothermal vents similar to those found on Earth. The presence of liquid water, coupled with potential energy sources and organic molecules, suggests environments potentially conducive to life, despite the extreme conditions of low temperature and darkness. The discovery of plumes erupting from Enceladus's surface, containing water vapor and organic molecules, further strengthens the case for investigating these icy worlds.

The search for life on Mars also benefits from our understanding of Earth's extremophiles. While the Martian surface is currently incredibly cold, dry, and exposed to high levels of radiation, evidence suggests that liquid water once flowed on the planet's surface, and possibly still exists in subsurface reservoirs. This raises the possibility of past or present microbial life, possibly adapted to the harsh Martian conditions. Future missions to Mars will focus on searching for evidence of past or present life, including the exploration of subsurface environments, the analysis of soil and rock samples for organic molecules, and the detection of biosignatures. The understanding gained from studying extremophiles on Earth informs the design of these missions and the interpretation of their results. The potential for past life on Mars is heightened by the discovery of organic molecules in Martian meteorites and the detection of methane in the Martian atmosphere, although the origin of these organic molecules and methane remains subject to further investigation.

Beyond our solar system, the search for extraterrestrial life is rapidly advancing with the discovery of thousands of exoplanets, planets orbiting stars other than our Sun. The study of these exoplanets, coupled with the knowledge gained from Earth's extremophiles, is helping to redefine our understanding of habitability. We are now identifying exoplanets within the "habitable zone" of their stars, where liquid water could exist on the surface. However, the definition of the "habitable zone" is becoming increasingly nuanced, as we recognize the possibility of subsurface oceans, like those found on Europa and Enceladus, and the existence of life in extreme environments on Earth.

The chemistry of life on Earth, based on carbon and water, provides a valuable framework for searching for life elsewhere. However, we should remain open to the possibility of life based on different chemistries, especially given the extreme environments in which life thrives on our own planet. For instance, some scientists propose that alternative biochemistries, such as those based on silicon or ammonia, might be possible under different environmental conditions. These alternative chemistries could lead to life forms with drastically different characteristics and metabolisms than those found on Earth. The search for extraterrestrial life therefore needs to be broad enough to encompass a wide range of possibilities, both in terms of environment and biochemistry.

The development of advanced technologies also plays a crucial role in the search for extraterrestrial life. Powerful telescopes, such as the James Webb Space Telescope, enable the detection and characterization of exoplanets, allowing us to study their atmospheres for signs of biosignatures, such as oxygen, methane, or other gases that might indicate the presence of life. Sophisticated robotic missions to other planets and moons, like those planned for Mars and the icy moons of the outer solar system, will further enhance our ability to search for evidence of past or present life. The combination of ground-based and space-based observations, coupled with advanced robotic exploration, represents a powerful toolset for the search for extraterrestrial life.

Finally, the study of the origin of life on Earth also provides crucial insights into the potential prevalence of life in the universe. If life arose relatively easily on Earth, despite the challenging conditions of early Earth, then it might be a common phenomenon throughout the universe. Conversely, if the origin of life proved to be an exceptionally rare event, then the likelihood of finding life elsewhere might be considerably lower. The search for extraterrestrial life is not merely a scientific endeavor; it is also a philosophical one, exploring our place in the cosmos and our potential for connection with other forms of life. The implications of discovering extraterrestrial life would be profound and far-reaching, impacting our understanding of biology, cosmology, and our place in the universe. The continued exploration of this area, through rigorous scientific research and technological advancements, is both thrilling and essential for expanding our knowledge and understanding of our universe and our place within it. The ongoing research in this field promises to unveil further secrets about the emergence and distribution of life, providing answers to questions that have captivated humanity for centuries. The pursuit of answers to these questions represents a collective endeavor, blending scientific rigor with a profound sense of wonder and curiosity about the universe.

The search for extraterrestrial life hinges on our ability to detect biosignatures – chemical or physical indicators that suggest the presence of life, past or present. Unlike searching for little green men, the realistic approach focuses on identifying subtle clues within the vastness of space. These biosignatures can be incredibly diverse, ranging from simple organic molecules to complex biological structures, and their detection requires a multifaceted approach combining sophisticated instrumentation, advanced data analysis techniques, and a deep understanding of the chemistry of life as we know it – and perhaps, as we don't yet know it.

One of the most promising approaches involves the analysis of planetary atmospheres. Powerful telescopes, like the James Webb Space Telescope (JWST), possess the capability to analyze the light passing through exoplanet atmospheres. By studying the absorption and emission spectra of this light, we can identify the presence of various gases. Certain gases, like oxygen (O2), methane (CH4), and nitrous oxide (N2O), are produced in significant quantities by biological processes on Earth, and their detection in the atmosphere of an exoplanet could suggest the presence of life. However, it's crucial to remember that these gases can also have abiotic origins, so their detection alone doesn't definitively prove life. The presence of multiple biosignature gases, in unusual combinations, or in quantities inconsistent with known geological processes, significantly strengthens the case for biological activity.

The detection of atmospheric biosignatures is complicated by several factors. The faintness of the signals from distant exoplanets necessitates extremely sensitive instrumentation. Furthermore, the signals can be easily masked or distorted by the light from the host star, making accurate measurements challenging. Sophisticated data analysis techniques are therefore required to separate the exoplanetary signal from the stellar background noise. Another significant challenge is the potential for false positives. Geological or chemical processes can produce gases that mimic biosignatures, leading to misinterpretations. Careful consideration of all possible abiotic sources is crucial for ensuring accurate interpretation of the data.

Beyond atmospheric analysis, the search for biosignatures also extends to the surfaces of planets and moons. Robotic missions, like the rovers currently operating on Mars, carry instruments capable of analyzing soil and rock samples for organic molecules – the building blocks of life. These instruments are designed to detect specific organic compounds, such as amino acids and nucleotides, that could indicate the presence of past or present life. The challenge here lies in distinguishing between organic molecules produced by biological processes and those formed through abiotic processes, such as meteor impacts or volcanic activity. Again, careful analysis and comparison with background levels are crucial. Furthermore, the harsh environmental conditions on some celestial bodies can lead to the degradation of organic molecules, making their detection even more difficult.

The search for biosignatures isn't limited to organic molecules. Certain isotopic ratios can also serve as indicators of life. Isotopes are variants of an element with different numbers of neutrons. Living organisms tend to preferentially utilize certain isotopes over others during metabolic processes, resulting in specific isotopic ratios in their biomass. The detection of these characteristic isotopic ratios in extraterrestrial samples could provide strong evidence for life. For instance, the relative abundance of carbon-12 versus carbon-13 can vary between biotic and abiotic sources. However, similar to organic molecule detection, the interpretation of isotopic ratios requires careful consideration of abiotic processes that could also influence isotopic fractionation.

Another fascinating avenue in the search for biosignatures involves the detection of unique biological structures. Microbial fossils, or even extant microorganisms, could leave behind distinctive traces in rocks or ice. However, identifying these structures requires highly sophisticated microscopy and imaging techniques. Furthermore, the preservation of biological structures over geological timescales can be highly challenging, and degradation can erase subtle features, making identification difficult.

The study of extremophiles on Earth plays a vital role in shaping the search for extraterrestrial life. By studying the diverse chemistries and survival strategies of organisms thriving in extreme conditions, we gain a broader understanding of the potential limits of life. This understanding informs the design of instruments and the interpretation of data obtained from extraterrestrial environments. It expands our definition of a "habitable zone" to encompass a much wider range of environments previously considered inhospitable.

For example, the discovery of extremophiles in highly acidic or alkaline environments on Earth expands the search for life to planets or moons with similar geochemical conditions. The identification of extremophiles capable of surviving intense radiation extends our search to celestial bodies previously thought too hostile to support life. The study of extremophiles that use alternative energy sources, like chemosynthesis, broadens our search to environments devoid of sunlight, such as deep subsurface oceans.

However, the search for life beyond Earth should remain open-ended and acknowledge that life elsewhere might not resemble life on Earth. The biochemistries of extraterrestrial life forms could differ significantly, making the detection of traditional biosignatures challenging. We should consider the possibility of alternative biochemistries based on elements other than carbon, like silicon, or solvents other than water, like ammonia. Such life forms might leave behind very different biosignatures, requiring us to broaden our search strategies.

The search for extraterrestrial biosignatures is a complex and challenging endeavor, demanding rigorous scientific methods and cutting-edge technologies. As our instrumentation improves and our understanding of life's diversity expands, the probability of detecting evidence of life beyond Earth becomes increasingly promising. While the definitive proof of extraterrestrial life remains elusive, the ongoing search continues to push the boundaries of our knowledge and inspires a sense of wonder about our place in the cosmos. The journey itself, with its technological advancements and scientific discoveries, represents a remarkable human endeavor, echoing our innate curiosity and our relentless pursuit of understanding the universe we inhabit.

| 4 |

Searching for Extraterrestrial Life

Our quest to understand the prevalence of life beyond Earth is inextricably linked to our ability to find and study planets orbiting other stars – exoplanets. The sheer number of stars in our galaxy, and indeed the universe, suggests a potentially vast population of exoplanets, many of which might harbor conditions suitable for life. However, detecting these planets is a monumental challenge, given their minuscule size and faint light compared to their host stars. Fortunately, astronomers have developed ingenious techniques to overcome this challenge, employing a range of observational methods to detect and characterize exoplanets.

One of the most successful methods for detecting exoplanets is the transit method. This technique relies on the subtle dimming of a star's light as a planet passes in front of it, from our perspective on Earth. Imagine a tiny speck momentarily eclipsing a much larger light source; this is the essence of the transit method. As the planet transits, a small but measurable decrease in the star's brightness occurs. The depth and duration of this dimming reveal information about the planet's size and orbital period. Kepler and TESS (Transiting Exoplanet Survey Satellite), two highly successful space telescopes, have employed the transit method to discover thousands of exoplanets. Their success stems from their ability to continuously monitor vast numbers of stars for periods of years, increasing the likelihood of detecting these infrequent transit events. The data collected by these missions is then subjected to rigorous analysis to confirm the planetary nature of the detected signals, carefully distinguishing genuine planetary transits from other sources of stellar variability.

The radial velocity method, also known as the Doppler spectroscopy method, provides a complementary approach to exoplanet detection. This method exploits the subtle gravitational tug exerted by a planet on its host star. As a planet orbits a star, its gravity causes the star to wobble slightly back and forth. This wobble induces a small Doppler shift in the star's light, causing the light to become slightly bluer when the star moves towards us and slightly redder when it moves away. By precisely measuring these minute changes in the star's spectral lines, astronomers can infer the presence of orbiting planets, and even estimate their mass. This method is particularly sensitive to detecting massive planets that orbit close to their stars, though advancements in instrumentation allow detection of smaller, farther-out planets as well. The radial velocity method, while less prolific in terms of exoplanet discoveries compared to the transit method, provides crucial information about the planets' masses and orbital characteristics, often in conjunction with data from the transit method. Combining data from both methods allows for a more complete characterization of the exoplanet, revealing its size, mass, and density.

While the transit and radial velocity methods have been remarkably successful, they are indirect detection methods. They infer the presence of planets based on their effects on the host star, rather than directly observing the planet itself. Direct imaging of exoplanets, in contrast, involves directly observing the faint light emitted or reflected by the planet. This is the most challenging method due to the overwhelming brightness of the host star, which easily drowns out the planet's faint signal. However, technological advancements in adaptive optics and coronagraphy are enabling astronomers to overcome this challenge. Adaptive optics compensate for the blurring effects of the Earth's atmosphere, while coronagraphs block out the starlight, allowing for the detection of nearby, relatively large, and often young planets. Direct imaging offers the potential to directly study the atmospheres of exoplanets, providing invaluable insights into their composition and potential habitability. It also allows for the study of exoplanets at larger orbital distances from their stars, which are often not detectable with the transit or radial velocity methods.

The characterization of exoplanets goes beyond simply detecting their presence; it involves determining their physical properties such as mass, radius, density, atmospheric composition, and temperature. This characterization is crucial for assessing the potential habitability of these planets. For example, a planet's density provides clues about its composition; a rocky planet will have a higher density than a gas giant. The analysis of an exoplanet's atmosphere, often conducted using spectroscopy, reveals the presence of various gases, like water vapor, methane, carbon dioxide, and oxygen. The detection of certain gases, especially in specific combinations and abundances, can serve as a powerful biosignature, suggesting the potential presence of life. The temperature of an exoplanet is also a crucial factor in determining its habitability; temperatures too hot or too cold can render a planet uninhabitable.

Furthermore, the location of the exoplanet within its star's habitable zone is a critical consideration. The habitable zone, also known as the Goldilocks zone, is the region around a star where the temperature is just right for liquid water to exist on the surface of a planet. Liquid water is considered essential for life as we know it, making the habitable zone a prime target in the search for extraterrestrial life. However, the precise boundaries of the habitable zone depend on several factors, including the star's mass and luminosity. Planets orbiting cooler, less luminous stars will have a smaller habitable zone closer to the star than planets orbiting hotter, more luminous stars.

The study of exoplanets is an active and rapidly evolving field. New discoveries and advancements in detection and characterization techniques are constantly expanding our understanding of the diversity of planetary systems in our galaxy. The discovery of thousands of exoplanets has revealed a wide range of planetary sizes, masses, compositions, and orbital configurations, far exceeding the diversity of planets within our own solar system. This diversity emphasizes the vast potential for discovering planets with conditions suitable for life.

The search for extraterrestrial life is inherently intertwined with the study of exoplanets. While the direct detection of life remains a distant goal, the characterization of exoplanets provides crucial information about the potential for life to exist beyond our solar system. The detection of biosignatures in exoplanet atmospheres would be a groundbreaking discovery, marking a monumental leap forward in our understanding of life's prevalence in the universe. However, even the absence of detectable biosignatures doesn't necessarily rule out the presence of life, as life might exist in forms or environments that are undetectable with current technology.

The quest to find extraterrestrial life is a long-term endeavor that requires sustained investment in research and development of new technologies. The advancements in observational astronomy, particularly in the realms of exoplanet detection and characterization, are critical to this pursuit. Future missions, including larger and more sophisticated telescopes, both ground-based and space-based, promise to significantly enhance our capabilities to detect and characterize exoplanets, potentially revealing the existence of life beyond Earth. The ongoing search for extraterrestrial life is a testament to human curiosity and our relentless pursuit of knowledge, driving us to explore the vast unknown and push the boundaries of our understanding of the universe. The sheer scale of the undertaking, however, necessitates a collaborative, international approach, combining the resources and expertise of scientists, engineers, and institutions from around the globe. Only through such a concerted effort can we hope to address one of humanity's most profound and enduring questions: Are we alone?

The search for extraterrestrial life takes a significant leap forward with the analysis of exoplanet atmospheres. While detecting exoplanets themselves is a considerable achievement, the true potential for discovering life lies in understanding the composition of their atmospheres. This involves sophisticated techniques and technologies capable of discerning the faint spectral fingerprints of gases within these distant worlds. The primary goal is the detection of biosignatures – gases or combinations of gases that strongly suggest the presence of life.

One of the most sought-after biosignatures is molecular oxygen (O�). On Earth, oxygen is a byproduct of photosynthesis, a process carried out by plants and other photosynthetic organisms. The presence of significant amounts of free oxygen in an exoplanet's atmosphere could therefore indicate the existence of photosynthetic life, although abiotic processes can also produce oxygen. The detection of oxygen alone, however, is not definitive proof of life. Its presence must be considered alongside other factors, like the presence or absence of other gases. A high concentration of oxygen without other gases associated with biological activity would raise questions about its origin, potentially suggesting non-biological production mechanisms. For instance, certain geological processes could, under specific conditions, release oxygen, creating a false positive. Therefore, a comprehensive atmospheric analysis is crucial, considering the interplay of different gases and their potential sources.

Methane (CH�) is another intriguing biosignature. While methane can be produced through geological processes, such as volcanism, it is also a significant byproduct of biological activity, including both anaerobic (oxygen-free) and aerobic processes. On Earth, methanogenic archaea, microorganisms thriving in anaerobic environments, are major producers of methane. The detection of methane, particularly in conjunction with other gases like carbon dioxide (CO�), can provide compelling, albeit indirect, evidence for life. However, the detection of methane, similar to oxygen, alone is not sufficient proof of life. A thorough analysis of the atmospheric context is necessary to rule out abiotic origins. For example, the presence of significant amounts of methane without other biosignatures, or coupled with an atmosphere inconsistent with life-supporting conditions (high temperatures or radiation levels), would suggest a geological or other abiotic source.

Water vapor (H�O) is another critical component in the search for life. While water itself is not a biosignature in the same way as oxygen or methane, its presence in the atmosphere is a strong indicator of potential habitability. Liquid water is considered essential for life as we know it, and the detection of water vapor suggests the possibility of liquid water existing on the planet's surface or subsurface, although atmospheric water vapor does not guarantee the presence of liquid water on the surface. The abundance of water vapor, its isotopic ratios, and its distribution within the atmosphere, combined with other factors such as temperature and pressure, provide crucial constraints on a planet's potential for habitability and, by extension, the presence of life. A planet with abundant water vapor but extreme surface temperatures, for instance, might not be capable of supporting life as we understand it.

The detection of these biosignatures is a complex undertaking, requiring advanced spectroscopic techniques. Spectroscopy involves analyzing the light from an exoplanet's atmosphere, which interacts with the gases present. Different gases absorb and emit light at specific wavelengths. By observing the spectrum of light passing through an exoplanet's atmosphere, we can identify the presence and abundance of various gases. However, the signals from exoplanet atmospheres are extremely faint, dwarfed by the light from their host stars. This requires sophisticated instruments and data analysis techniques to separate the planetary signal from the stellar signal. Ground-based telescopes, while capable of achieving high spectral resolution, are limited by the Earth's atmosphere, which can distort and blur the light. Space-based telescopes, like the James Webb Space Telescope (JWST), offer a significant advantage because they are positioned above the Earth's atmosphere, providing clearer and more detailed observations.

The challenges involved in atmospheric analysis extend beyond the technical limitations of current technology. The interpretation of data is a complex process requiring careful consideration of various factors. For example, the presence of a biosignature might be masked by other atmospheric phenomena, or the biosignature itself may have abiotic origins. Moreover, the diversity of life on Earth, ranging from extremophiles thriving in extreme environments to complex multicellular organisms, suggests that extraterrestrial life might manifest in ways that are currently beyond our ability to detect. We may simply be searching for the wrong biosignatures or looking in the wrong places.

Another significant challenge relates to the distance to exoplanets. The light from even the closest exoplanets is incredibly faint, requiring extremely long integration times to accumulate sufficient signal for spectral analysis. This can limit the amount of data that can be obtained and result in large uncertainties in the detected abundances of atmospheric constituents. Furthermore, the characterization of an exoplanet's atmosphere typically requires multiple observations, spanning different phases of the planet's orbit and taking advantage of various techniques to minimize observational noise and improve signal-to-noise ratio. These multiple observations further extend the time commitment for a thorough characterization.

Despite these challenges, the search for biosignatures in exoplanet atmospheres is a rapidly advancing field. Technological advancements in telescope design, detector sensitivity, and data analysis techniques are continuously pushing the boundaries of what is possible. Larger and more powerful telescopes, both ground-based and space-based, are being planned and built, enabling more sensitive and detailed observations. Advances in computational techniques are also aiding in data analysis, enabling the extraction of meaningful information from increasingly complex datasets. The development of sophisticated atmospheric models is further helping to interpret the observed data and constrain the potential habitability of exoplanets.

The future of exoplanet atmospheric analysis is bright. Future missions, such as the Extremely Large Telescope (ELT) and the Nancy Grace Roman Space Telescope, will provide unprecedented capabilities to detect and characterize exoplanet atmospheres. These next-generation telescopes will boast significantly larger collecting areas than current telescopes, dramatically improving their sensitivity and ability to observe fainter signals. They will also incorporate advanced adaptive optics and coronagraphy systems, further enhancing their ability to eliminate the glare of host stars and isolate the signal from exoplanet atmospheres. These technological advancements, coupled with continued progress in data analysis and theoretical modeling, promise to significantly enhance our ability to detect and interpret biosignatures in exoplanet atmospheres. The path forward necessitates a multidisciplinary approach, integrating expertise from various fields such as astronomy, astrobiology, atmospheric science, chemistry, and computer science. This collaborative effort will be essential to overcome the many challenges in this rapidly evolving field and advance the quest for life beyond Earth. The detection of definitive biosignatures would be a monumental discovery, profoundly altering our understanding of our place in the universe. Even the absence of detectable biosignatures will contribute to our understanding of the factors that influence the emergence and evolution of life throughout the cosmos. The search continues, and the discoveries made along the way are crucial, regardless of the ultimate outcome.

The search for extraterrestrial life extends beyond the analysis of exoplanet atmospheres; it also encompasses the ambitious quest to detect extraterrestrial intelligence – the realm of SETI, the Search for Extraterrestrial Intelligence. Unlike the search for biosignatures, which focuses on indirect evidence of life, SETI directly seeks out signals of technological civilizations. This involves listening for artificial radio waves, laser pulses, or other forms of communication that might originate from beyond our solar system.

The fundamental assumption underlying SETI is that other intelligent civilizations exist and have developed technologies capable of interstellar communication. This assumption, while not proven, is rooted in the sheer vastness of the universe and the probability, however small, that life – and intelligent life – might arise elsewhere. The sheer number of stars and planets in the observable universe is staggering – trillions upon trillions – making the possibility of Earth being unique seem increasingly improbable. The Drake Equation, while often debated, serves as a useful framework for conceptualizing the number of potentially detectable civilizations in our galaxy. While many of the equation's factors remain highly uncertain, it highlights the critical variables involved, including the rate of star formation, the fraction of stars with planets, the fraction of planets suitable for life, and the fraction of those planets that actually develop intelligent life capable of interstellar communication.

SETI's methods are primarily observational. Early SETI projects, like Project Ozma in the 1960s, used radio telescopes to scan the sky for narrowband signals – signals concentrated within a very narrow frequency range, a characteristic often associated with artificial transmissions. The rationale behind focusing on narrowband signals is that natural astrophysical phenomena tend to produce broadband emissions spanning a wider range of frequencies. A narrowband signal, therefore, would stand out as an anomaly, potentially indicating an artificial origin. These early efforts were limited by the sensitivity and capabilities of available technology, scanning only a small fraction of the sky and a limited range of frequencies.

Modern SETI projects employ far more sophisticated techniques and technology. The Allen Telescope Array (ATA), for example, is a large array of radio telescopes designed to search for signals from extraterrestrial civilizations. Its large collecting area and ability to observe many frequencies simultaneously significantly increase its sensitivity and the area of the sky it can scan. The ATA utilizes advanced signal processing techniques to sift through the vast amount of data collected, filtering out terrestrial interference and identifying potential candidate signals for further analysis.

Beyond radio waves, SETI also explores the potential of detecting optical signals – laser pulses. Laser beams can transmit information over interstellar distances with considerably higher data rates compared to radio waves. However, detecting optical signals requires extremely sensitive detectors capable of distinguishing faint laser pulses from background noise, a significantly more challenging task than detecting radio signals. Current efforts are focused on developing these advanced optical detection systems and implementing search strategies optimized for this specific approach.

The challenges faced by SETI are immense. The sheer volume of data generated by SETI projects is enormous, requiring advanced computational techniques to analyze and filter effectively. Distinguishing artificial signals from natural radio emissions or terrestrial interference is a major hurdle. The potential for false positives – signals that appear to be extraterrestrial but ultimately have terrestrial origins – is ever-present, and robust verification techniques are crucial to minimize misinterpretations. The distances involved pose another significant challenge. The faintness of signals originating from other star systems necessitates sensitive detection equipment and long observation times, often spanning many years.

Furthermore, the nature of potential extraterrestrial signals remains uncertain. We are essentially searching for something we don't know how to definitively identify. Extraterrestrial civilizations might employ communication technologies far beyond our current comprehension. They might utilize methods or frequencies we haven't even considered, making our search methods potentially incomplete. Moreover, the very notion of communication itself might differ significantly from our anthropocentric perspective. Extraterrestrial intelligence might not wish to communicate with us, or their means of communication might not align with our expectations. This raises the profound question of whether we are looking for the right kinds of signals, in the right places, and using the right methods.

The philosophical implications of detecting extraterrestrial intelligence are profound. Such a discovery would revolutionize our understanding of our place in the universe, forcing us to reconsider our assumptions about life and intelligence. It would have significant implications for science, philosophy, religion, and possibly even our political and societal structures. The impact on humanity's collective consciousness would be immeasurable. The discovery of another intelligent civilization would not only expand our scientific knowledge but also challenge our fundamental worldview.

Despite the immense challenges, SETI remains a compelling and important endeavor. The potential rewards far outweigh the risks, justifying the substantial investment of resources and time. The search continues, fueled by scientific curiosity, the desire to understand our place in the cosmos, and the hope of making a momentous discovery that could forever alter humanity's perception of itself and the universe. Even the absence of detection, after exhaustive searches, provides valuable insights into the prevalence (or lack thereof) of technologically advanced civilizations in the galaxy and informs further research strategies. The quest for extraterrestrial intelligence is not merely a scientific pursuit; it's a testament to humanity's innate curiosity, our relentless pursuit of knowledge, and our enduring hope of not being alone in the vast expanse of the universe. The search for extraterrestrial intelligence is a journey of discovery, a testament to human ingenuity, and a profound exploration of our place within the cosmos. The quest continues, driven by an unwavering belief that the universe may harbor life beyond our wildest imaginations, life that could reshape our understanding of existence itself. The potential discovery of extraterrestrial intelligence would redefine humanity's role in the universe, impacting our scientific, philosophical, religious, and social perspectives. This grand endeavor is a testament to our intellectual curiosity and our unyielding pursuit of knowledge.

The search for extraterrestrial life isn't confined to passive observation from Earth; it also involves actively exploring other celestial bodies that might harbor life, past or present. This exploration relies heavily on robotic missions, offering a cost-effective and relatively safe way to investigate environments too hostile for human astronauts. These robotic explorers serve as our eyes, ears, and hands in the cosmos, gathering data and samples that significantly advance our understanding of planetary evolution and the potential for life beyond Earth.

The exploration of Mars stands as a prime example of robotic missions' crucial role in astrobiology. Decades of robotic missions, beginning with the Mariner and Viking missions in the 1970s, have progressively revealed a Mars far more complex than initially imagined. While the Viking landers failed to find definitive evidence of extant life, they significantly improved our understanding of Martian geology and climate, laying the groundwork for future missions. Subsequent missions like the Mars Pathfinder, Spirit, Opportunity, Curiosity, and Perseverance rovers have dramatically increased our knowledge.

These rovers have traversed the Martian surface, analyzing rocks and soil, searching for evidence of past water activity, and even collecting samples for potential future return to Earth. The Perseverance rover, in particular, is equipped with sophisticated instruments to detect potential biosignatures – chemical indicators of past or present life – and is actively collecting rock and soil samples for eventual retrieval by a future mission. The analysis of these samples holds immense potential for unveiling evidence of past microbial life on Mars, potentially revolutionizing our understanding of the prevalence of life in the universe.

Beyond the surface exploration, orbital missions play a vital role in gaining a broader perspective of Mars. Orbiters, such as the Mars Reconnaissance Orbiter (MRO) and the Mars Odyssey, provide high-resolution imagery, mapping the Martian surface in unprecedented detail and identifying potential landing sites for future missions. They also study the Martian atmosphere, climate, and geological processes, giving context to the findings of surface missions. These orbital observations are crucial for understanding the planet's past habitability and identifying regions that may still harbor subsurface water or other life-sustaining resources. The synergy between orbital and surface missions is essential for a comprehensive understanding of the Red Planet's potential to support life.

While Mars is a focal point, other celestial bodies also hold immense potential for harboring extraterrestrial life. Europa, one of Jupiter's four Galilean moons, is a prime candidate. Underneath its icy surface, Europa is believed to possess a vast subsurface ocean of liquid water, possibly containing more water than all of Earth's oceans combined. The presence of this subsurface ocean, coupled with evidence suggesting hydrothermal activity on Europa's seafloor, makes it a highly compelling target in the search for extraterrestrial life. Future robotic missions to Europa, such as the Europa Clipper mission, will study the moon's surface and subsurface ocean, attempting to determine its composition and assess its potential habitability. The Clipper mission will use a suite of sophisticated instruments to study the moon's surface, including imaging spectrometers to map the surface composition, radar instruments to penetrate the ice sheet and probe the subsurface ocean, and magnetometers to study Europa's magnetic field and infer the presence of a subsurface ocean. The goal is to determine whether Europa's ocean could sustain life and, ultimately, to determine if life might exist there.

Similar to Europa, other icy moons within our solar system, such as Enceladus (Saturn's moon) and Titan (Saturn's largest moon), are also considered promising candidates for harboring life. Enceladus has been shown to have geysers erupting water vapor and ice particles from its subsurface ocean, offering a unique opportunity to sample the ocean's contents without the need to drill through the ice sheet. Titan, with its hydrocarbon lakes and seas, represents a unique environment potentially capable of hosting life forms very different from those on Earth. Future missions are planned to further explore these icy worlds, continuing the search for evidence of life beyond Earth.

The design and engineering of these robotic missions are marvels of technological innovation. Robotic spacecraft must withstand the harsh conditions of space, including extreme temperatures, radiation, and micrometeoroid impacts. They require sophisticated autonomous navigation systems to operate effectively in remote locations, often with limited or delayed communication with Earth. The instruments carried by these spacecraft are also highly advanced, capable of conducting complex scientific analyses and collecting high-quality data under challenging conditions. Miniaturization, power efficiency, and redundancy are key design considerations to ensure the robustness and longevity of these robotic explorers.

Furthermore, the data analysis involved in robotic space exploration is a significant undertaking. The enormous amounts of data generated by these missions require sophisticated computational techniques and algorithms to process, interpret, and extract meaningful scientific insights. Artificial intelligence (AI) and machine learning (ML) are increasingly being utilized to automate data analysis tasks, allowing scientists to focus on higher-level scientific interpretations and hypotheses. The combination of advanced robotics and AI is accelerating the pace of discovery in astrobiology.

However, challenges remain. The development and launch of robotic missions are expensive and time-consuming, requiring significant investments of resources and expertise. The long travel times to other celestial bodies present significant logistical challenges, demanding reliable and durable spacecraft capable of operating autonomously for extended periods. Data transmission from distant locations can be slow and limited, potentially hindering real-time analysis and decision-making. Despite these challenges, the potential rewards of discovering extraterrestrial life far outweigh the difficulties, justifying continued investment in robotic missions and the development of increasingly sophisticated technologies.

The future of robotic exploration in the search for extraterrestrial life looks bright. More ambitious missions are planned to explore a broader range of celestial bodies, employing even more advanced technologies. The development of more robust, autonomous, and adaptable robots will allow for more in-depth exploration of challenging environments. The utilization of AI and machine learning will further enhance data analysis capabilities, accelerating the pace of discovery. The integration of multiple spacecraft and instruments will provide a more holistic view of planetary systems and potentially increase the chances of detecting biosignatures. The relentless pursuit of scientific knowledge, coupled with technological advancements, ensures that the search for extraterrestrial life using robotic missions will continue to be a major focus of astrobiological research for decades to come. The success of these robotic missions hinges on a combination of innovative engineering, sophisticated data analysis techniques, and a deep understanding of planetary science and astrobiology. The quest for extraterrestrial life is a grand endeavor, a testament to human curiosity and our relentless pursuit of knowledge in the vast expanse of the universe.

Despite the remarkable progress in robotic exploration and data analysis, significant technological limitations still hinder our search for extraterrestrial life. These limitations encompass several key areas: instrumentation sensitivity, power constraints, communication delays, and data processing capacity. Overcoming these hurdles requires a concerted effort in technological innovation, pushing the boundaries of current capabilities to enhance our ability to detect and characterize life beyond Earth.

One of the most significant limitations is the sensitivity of our current instruments. Detecting biosignatures – the chemical fingerprints of life – requires incredibly sensitive instruments capable of identifying minute traces of organic molecules or other indicators of biological activity. While current instruments aboard missions like Perseverance are impressive, they are still limited in their ability to detect extremely faint signals or differentiate between abiotic (non-biological) and biotic (biological) processes. For example, the detection of methane on Mars has sparked considerable excitement, as methane can be produced by both biological and geological processes. Distinguishing between these two sources requires highly sophisticated analytical techniques that can accurately quantify isotopic ratios and other characteristics of the methane, and even then, the interpretation is not straightforward. Future missions will need instruments with even greater sensitivity and specificity to unequivocally identify biosignatures in the face of complex geochemical processes. This necessitates the development of novel sensor technologies, such as advanced mass spectrometers with higher resolution and sensitivity, and more refined techniques for analyzing the isotopic composition of organic molecules.

Power constraints pose another significant challenge. Robotic missions are typically limited by the amount of power they can generate, usually through solar panels or radioisotope thermoelectric generators (RTGs). This limited power supply restricts the operational time and capabilities of instruments, particularly those that require high energy consumption. For example, analyzing samples using highly sensitive techniques like gas chromatography-mass spectrometry (GC-MS) demands considerable power, limiting the number of samples that can be analyzed during a mission's lifespan. Advancements in power generation technologies are crucial to overcome these limitations. The development of more efficient solar panels, improved RTGs, and potentially new power sources like nuclear fission reactors could significantly enhance the capabilities of future robotic missions. Miniaturization of instruments also plays a crucial role, as smaller, more energy-efficient instruments can conserve power and enable more extensive exploration.

The vast distances to other celestial bodies necessitate long travel times, leading to considerable communication delays. The time it takes for a signal to travel from Earth to a distant spacecraft and back can range from minutes to hours, depending on the distance. This delay significantly complicates the operation of robotic missions, as real-time control and decision-making are severely restricted. Autonomous navigation and decision-making capabilities are critical to mitigate these delays, and advanced AI and machine learning algorithms are crucial in this regard. These algorithms allow spacecraft to operate more independently, adapting to unforeseen circumstances and making decisions without requiring immediate instructions from Earth. However, even with advanced autonomy, the lack of real-time communication still presents a significant limitation. Furthermore, the bandwidth limitations of communication systems can constrain the volume and speed of data transmission, leading to potential bottlenecks in the scientific analysis of acquired data. Developments in advanced communication technologies, such as high-bandwidth laser communication systems, are essential to overcome these limitations and enable more efficient and timely data transfer.

Data processing and analysis represent another significant hurdle. Robotic missions generate massive amounts of data, far exceeding the capacity of current data processing infrastructure. Analyzing this data efficiently and effectively requires advanced computational techniques and algorithms. While AI and machine learning are increasingly used to automate aspects of data analysis, the sheer volume of data and the complexity of the analysis needed remain significant challenges. Developing more efficient algorithms and more powerful computing systems, including cloud-based computing and distributed computing platforms, is essential to process and interpret the data efficiently. Furthermore, the development of automated data analysis pipelines that can handle diverse data types and identify relevant features is also critical. The human element remains crucial in interpreting the findings of these automated analyses; scientists are needed to integrate diverse datasets, validate findings, and formulate new hypotheses.

Moreover, the challenges extend beyond the technological sphere into the realm of scientific understanding. Our current understanding of extremophiles – organisms that thrive in extreme environments on Earth – informs our search for life beyond Earth. However, the conditions on other celestial bodies may be so different from anything found on Earth that our existing paradigms of life may be insufficient. The possibility of life based on different chemistries or utilizing energy sources unknown to us presents a significant scientific challenge. The development of new theoretical frameworks and experimental approaches is vital for expanding our understanding of the possible forms that extraterrestrial life may take. This will require interdisciplinary collaborations, merging expertise from fields like astrobiology, chemistry, physics, and computer science.

The limitations discussed above highlight the urgency for continued technological innovation in our pursuit of extraterrestrial life. Investments in research and development are crucial for overcoming these hurdles and enhancing our ability to detect and characterize life beyond Earth. Future missions will benefit from more robust, adaptable, and energy-efficient spacecraft and instruments. The development of new sensor technologies capable of detecting a wider range of biosignatures and operating under more extreme conditions is essential. Advanced communication systems will mitigate the limitations of distance and enable faster data transfer. Finally, significant advancements in data processing, analysis, and interpretation will enable scientists to efficiently extract meaningful insights from the vast amounts of data generated by future missions. The search for extraterrestrial life remains a grand challenge, but through sustained technological advancement and collaborative scientific efforts, we can significantly enhance our ability to answer one of humanity's most fundamental questions. The potential rewards of this pursuit far outweigh the significant investment required, offering the transformative potential to fundamentally reshape our understanding of life, the universe, and our place within it.

| 5 |

Analyzing Reported Alien Abduction Accounts

The sheer volume of reported alien abduction accounts, spanning decades and continents, presents a fascinating, if controversial, area of study. While the scientific community largely remains skeptical, the persistent and consistent nature of these reports demands a considered analysis. Ignoring them entirely risks overlooking potential sociological insights, even if the claims of extraterrestrial encounters themselves lack definitive proof. Our focus here isn't to validate or invalidate these accounts individually, but to examine their commonalities, explore potential psychological and sociological interpretations, and highlight the challenges involved in scientifically evaluating them.

A common thread running through many alien abduction narratives is the experience of being taken aboard a craft, often described as metallic, sleek, and technologically advanced beyond anything currently known to humankind.

These crafts are frequently portrayed as silently hovering or moving with impossible speed and agility. Descriptions vary widely, but common features include bright lights, unusual sounds, and a sense of overwhelming powerlessness on the part of the abductee. The interior of the craft is often described as sterile, brightly lit, and containing sophisticated-looking equipment – often described in ways inconsistent with our current understanding of physics and engineering.

The individuals involved in these reported abductions describe a range of physical examinations or procedures, often involving the insertion of needles or probes, taking of bodily samples, or the implantation of devices. These descriptions are strikingly similar across different accounts, despite the lack of any known communication between the individuals reporting these experiences. The details of these procedures vary, but the recurring theme of intrusive medical examinations is a prominent feature. The emotional impact of these experiences is consistently described as intense fear, confusion, and a profound sense of violation. Furthermore, there's often a sense of memory alteration or gaps in recall surrounding the abduction event itself, further complicating attempts at objective evaluation.

Beyond the physical aspects, many abductees report encounters with the beings themselves. Descriptions often depict small, grey-skinned humanoids with large heads, black eyes, and elongated limbs, though accounts also include other types of entities. Interactions are typically described as non-verbal, sometimes involving telepathic communication or the impression of being controlled. The purpose of these interactions remains unclear within the narratives, with some abductees reporting feelings of being studied, experimented upon, or even seemingly communicated with on an emotional or spiritual level. Such encounters often lead to lasting psychological effects, ranging from anxiety and depression to profound changes in worldview and belief systems.

The psychological impact of reported abductions is a crucial area of study. Many researchers have explored the possibility that these experiences are rooted in sleep paralysis, hypnagogic hallucinations, or other psychological phenomena. Stress, trauma, or underlying mental health conditions could also contribute to the creation and recall of these experiences. However, this doesn't negate the subjective reality of the experiences for those who report them. Many abductees undergo intense emotional distress and seek therapeutic support to cope with the psychological aftermath. The fact that many individuals independently report extremely similar experiences suggests the need for investigation beyond simple dismissal as individual psychological breakdowns. This requires nuanced research considering both individual psychology and broader social influences.

The sociological significance of these reports is equally important. The prevalence of abduction narratives might reflect a cultural fascination with extraterrestrial life, a response to anxieties about technological advancements, or a manifestation of underlying social anxieties about control and power. The widespread media coverage of UFOs and alien encounters has undoubtedly shaped public perception and influenced the content of these reports. The narratives often reflect contemporary anxieties, integrating elements of science fiction, religious beliefs, and societal fears into the experience. Analyzing the narratives within their sociocultural context offers insights into the collective imagination and cultural anxieties of the time in which they emerged. Furthermore, the patterns and trends observed in abduction reports over time can provide valuable sociological data on the evolution of societal anxieties and beliefs.

Another crucial aspect to consider is the potential for misidentification, misinterpretation, or confabulation. Sleep disturbances, vivid dreams, unusual weather phenomena, or encounters with military aircraft or other technologies could be easily misinterpreted as alien encounters, particularly under conditions of stress or heightened suggestibility. The act of recalling and describing the experience, especially under the influence of leading questions or suggestion, can also distort or embellish the memory. This is where the rigorous methodology of scientific research is crucial; to differentiate between genuine encounters (if any exist) and experiences rooted in psychological or other earthly explanations.

The challenge in studying alien abduction reports lies in the inherent difficulty of obtaining verifiable evidence. The lack of physical evidence, the unreliability of eyewitness testimony, and the often fragmented and altered memories make it incredibly difficult to draw definitive conclusions. Furthermore, the phenomenon often resists standard scientific investigation methods, due to its subjective and elusive nature. Researchers must critically analyze accounts, controlling for biases, considering potential alternative explanations, and searching for patterns and consistencies across a wide range of narratives. This requires interdisciplinary collaboration, drawing on expertise from psychology, sociology, anthropology, and other relevant fields.

Moreover, the lack of cooperation from official agencies and the secrecy surrounding some investigations further hinders scientific scrutiny. While the gradual declassification of some government documents related to UFO sightings is a positive step, significant barriers remain. This lack of transparency breeds suspicion and makes it difficult for researchers to fully analyze the available data. The challenge, therefore, is to find a middle ground between accepting every report as factual and dismissing them all as delusion or fabrication. The cautious approach involves acknowledging the subjective nature of the experiences while carefully examining the common themes, patterns, and the wider social context.

While we cannot definitively conclude whether these reports represent genuine extraterrestrial encounters, the persistent and consistent nature of the accounts, coupled with their profound psychological and sociological impact, warrants further investigation. By using a multidisciplinary approach that integrates psychological, sociological, and anthropological perspectives, we can gain a more comprehensive understanding of the phenomenon. Such research is crucial to understanding the reported experiences themselves, as well as the impact on the individuals reporting them. It also helps us better understand the complex interplay between individual experiences, cultural narratives, and collective anxieties about our place in the universe. The study of alien abduction reports, therefore, offers a unique lens through which to examine the human condition and our fascination with the unknown, even if it doesn't offer a definitive answer to the question of extraterrestrial visitation. The enduring mystery itself drives the need for continued, rigorous research. Perhaps, through this process, we may uncover valuable insights into the human psyche, the power of belief, and the deep-seated human longing to connect with something beyond ourselves.

The prevalence of reported alien abduction experiences necessitates a thorough examination of the psychological and sociological underpinnings that might contribute to their creation and persistence. While the possibility of extraterrestrial contact remains a topic of scientific debate, the consistent themes and patterns within these accounts warrant exploration beyond the simple dismissal of individual psychological issues. This analysis delves into the complex interplay of individual suggestibility, cultural influences, mass hysteria, and societal anxieties, all of which can shape the interpretation and reporting of unusual experiences.

One crucial psychological factor is suggestibility, the degree to which an individual is susceptible to accepting and internalizing suggestions from external sources. Suggestibility varies widely among individuals, influenced by factors like personality traits, emotional state, and social context. In the context of alien abduction reports, suggestive questioning during hypnosis or therapeutic sessions might inadvertently lead to the creation or reinforcement of false memories. The power of suggestion can be particularly potent when dealing with individuals experiencing heightened anxiety or emotional distress, possibly contributing to the recall of elaborate but ultimately inaccurate experiences. Furthermore, the suggestive influence of media portrayals of alien abductions – through films, television shows, and books – cannot be overlooked. The consistent portrayal of specific alien types, spacecraft designs, and procedures may unconsciously shape the narratives of those who report such experiences, creating a framework upon which individuals structure their recalled experiences.

The role of sleep paralysis and hypnagogic hallucinations further complicates the picture. Sleep paralysis is a temporary inability to move or speak while falling asleep or waking up, often accompanied by vivid hallucinations. Hypnagogic hallucinations are vivid sensory experiences that occur during the transitional state between wakefulness and sleep. These phenomena, commonly experienced by a significant portion of the population, can be exceptionally realistic and terrifying, potentially leading to misinterpretations of normal physiological occurrences as encounters with extraterrestrial entities. The ambiguity of the liminal space between sleep and wakefulness provides fertile ground for the creation of unusual and emotionally powerful memories that might be misconstrued as alien abductions. The subjective experience is undeniably real for the individual, irrespective of the underlying neurological or physiological causes.

Cultural influences significantly shape the narrative framework through which individuals understand and report their experiences. The cultural myths and societal anxieties of the time period during which an individual experiences a potential abduction heavily influences how they frame their memory of the event. For example, reports from the Cold War era might reflect societal anxieties surrounding government surveillance and technological control, while reports from more recent decades could incorporate fears about genetic modification and the potential for unknown technologies to manipulate the human body. The cultural lens through which individuals interpret unusual experiences is often subconscious, coloring the recall and reconstruction of events to fit existing belief systems and social narratives.

Furthermore, the pervasive influence of mass media in disseminating information about UFOs and alien abductions plays a crucial role in shaping public perception. Media coverage, often sensationalized for dramatic effect, can normalize the concept of alien encounters and inadvertently provide a framework for individuals seeking to understand their own unusual experiences. This influence is particularly potent in situations where individuals feel uncertain about what they have experienced, potentially leading to a self-fulfilling prophecy where ambiguous experiences are interpreted through the pre-existing framework presented by the media.

The concept of mass hysteria, or collective psychogenic illness, also deserves consideration, especially when analyzing clusters of similar reports in a specific region or time period. Mass hysteria involves the rapid spread of physical or psychological symptoms within a group of individuals, often without any clear underlying medical or physical cause. The shared experience and emotional contagion inherent in mass hysteria can contribute to the amplification and reinforcement of individual experiences, leading to the homogenization of accounts and the creation of compelling but potentially inaccurate collective narratives. This phenomenon, though difficult to definitively prove in the context of alien abductions, suggests the importance of considering the social and emotional factors that might contribute to the spread and consistency of specific narratives.

Beyond these psychological and social elements, the potential for misidentification, misinterpretation, and outright confabulation must be acknowledged. Unusual weather phenomena, military exercises, and advanced technologies may be misconstrued as alien encounters, especially under conditions of stress or reduced visibility. The fallibility of human memory, including the potential for confabulation (the unintentional creation of false memories), further complicates attempts to objectively evaluate these reports. The reconstruction of memory is an active process, subject to the influence of social pressures, suggestibility, and the natural tendency to simplify and organize complex experiences. This highlights the need for rigorous methodology in assessing abduction accounts, recognizing the limitations of relying solely on subjective recall.

In conclusion, the analysis of alien abduction accounts requires a nuanced understanding of the complex interplay between individual psychology, social influences, and the potential for misinterpretation. While the possibility of extraterrestrial contact remains a compelling scientific question, a thorough examination of the psychological and sociological factors that might contribute to these reports is essential to achieving a comprehensive and objective understanding of the phenomenon. By employing a multidisciplinary approach that incorporates insights from psychology, sociology, anthropology, and other relevant fields, researchers can gain a deeper appreciation of both the subjective experiences of those reporting abductions and the broader social and cultural forces that shape their narratives. This holistic approach allows for a more robust assessment, avoiding the extremes of either fully accepting every account as factual or entirely dismissing them as delusions. The goal is not to resolve the question of extraterrestrial visitation definitively, but to deepen our understanding of the human experience, our perception of reality, and the power of belief in shaping our understanding of the world around us. The enduring mystery surrounding alien abduction reports serves as a catalyst for exploring fundamental aspects of the human condition, and through that exploration, perhaps we can uncover insights that extend far beyond the question of extraterrestrial visitors themselves.

The sheer volume of reported alien abduction accounts, despite their inherent subjectivity and lack of concrete physical evidence, reveals a fascinating tapestry of recurring themes and patterns. These common threads, while not definitive proof of extraterrestrial visitation, offer valuable insights into the human experience and the ways in which we process and interpret unusual or unexplained events. A systematic analysis of these recurring motifs, devoid of preconceived notions, is crucial to understanding the broader phenomenon.

One of the most prevalent themes involves the description of the abducting entities themselves. While variations exist, a common archetype emerges: humanoid figures, often described as having large heads, large black eyes, and gray or pale skin. These descriptions, while varying in detail across accounts, share striking similarities, leading some to propose a shared cultural template or a subconscious influence shaping the memories of those reporting these experiences. The consistency across different geographical locations and cultural backgrounds suggests that the image of the "Gray alien" might not necessarily reflect an actual extraterrestrial species but rather a product of collective imagination and cultural dissemination. This does not invalidate the experiences of individuals reporting these encounters, but it prompts a critical examination of how cultural narratives shape the recall and reporting of unusual experiences. The influence of popular culture, with its well-established imagery of alien abductions, undoubtedly contributes to this consistency. Movies, television shows, and books have played a significant role in establishing a widely recognized visual representation of extraterrestrials, potentially shaping the subconscious expectations and interpretations of individuals who report abductions.

The accounts often describe the abduction process itself in strikingly similar ways. Many individuals report being paralyzed, unable to move or cry out for help. This paralysis, often experienced during the initial encounter or during specific procedures described within the accounts, aligns with known phenomena like sleep paralysis, adding another layer of complexity to the analysis. Sleep paralysis, a common physiological experience, can involve vivid hallucinations and a sense of intense fear and helplessness. While not proof against the possibility of extraterrestrial encounters, it provides a potential explanation for some aspects of the reported experiences. Furthermore, the accounts frequently describe the use of advanced technology, often depicted as sophisticated medical instruments or diagnostic devices. These descriptions frequently include examinations of the abductee's body, often focusing on reproductive organs or blood samples. The consistency of these descriptions, again, raises questions about cultural influences and the potential role of subconscious expectations in shaping the narrative. It's crucial to distinguish between the subjective experience of the individuals and any objective assessment of the events' actual occurrence.

The alleged procedures performed during these abductions frequently center on themes of medical experimentation and genetic manipulation. Many accounts detail invasive procedures, often involving the extraction of body fluids or the implantation of devices. This recurrent motif raises significant questions about the underlying anxieties of our society. The fear of medical experimentation and the violation of bodily autonomy resonate with broader societal concerns surrounding genetic engineering and the ethical implications of manipulating human DNA. These anxieties can subconsciously influence the recall and interpretation of experiences, leading to the creation of narratives that align with prevailing cultural anxieties. This does not mean that the individuals reporting these experiences are consciously fabricating events, but rather that their memories are shaped and filtered through the lens of their own cultural and personal experiences.

Another recurring element in many accounts is the theme of communication, or rather, the lack thereof. While some individuals claim to have engaged in telepathic communication with the abductors, many others report a lack of verbal interaction, highlighting the ambiguity and inexplicability that often characterize these encounters. The absence of clear communication strengthens the case for examining alternative explanations, such as psychological or sociological factors, which might account for some aspects of the reported experiences. The difficulty in establishing objective verification for these encounters underscores the challenges in relying solely on subjective recall.

Furthermore, the temporal distortions reported by abductees are a fascinating and consistently recurring theme. Many individuals describe a subjective experience of time that differs from objective time, with hours or even days feeling like mere minutes or vice versa. This phenomenon mirrors experiences reported in other altered states of consciousness and aligns with the effects of extreme stress and fear on time perception. Moreover, the perceived lack of control and the overwhelming sense of powerlessness further contribute to the distortion of time perception. The subjective experience of time during highly stressful or traumatic events is well-documented, further suggesting alternative explanations for this recurring aspect of reported abductions.

The post-abduction experiences also reveal recurring patterns. Many individuals report significant psychological changes following the alleged encounter, including anxieties, fears, and a profound sense of disorientation. These psychological symptoms could be related to the traumatic nature of the reported experience, regardless of its actual cause. The stress and fear associated with an encounter interpreted as an abduction, even if based on a misinterpretation of a normal physiological event or a hallucination, are likely to have long-lasting psychological consequences. This is consistent with our understanding of trauma and its long-term effects.

The geographic distribution of reported abductions also raises interesting questions. While reports come from across the globe, certain areas seem to exhibit a higher concentration of reported cases. This geographic clustering might be attributed to several factors, including regional variations in media coverage, cultural beliefs, and even local environmental factors that could influence the frequency of sleep paralysis or other related phenomena. However, the absence of a clear correlation between geographic location and reported abduction rates necessitates further investigation. A comprehensive analysis must account for the possibility of biases in reporting, the influence of media sensationalism, and the lack of standardized methodology for collecting and analyzing these reports.

In conclusion, while definitive conclusions about the reality of alien abductions remain elusive, the consistent patterns and themes emerging from these accounts offer valuable insights into the human experience. The recurring motifs—the description of the abductors, the abduction process, the alleged procedures, communication issues, temporal distortions, and post-abduction psychological changes—reveal a compelling blend of psychological, sociological, and potentially physiological factors. A balanced approach demands considering alternative explanations, such as sleep paralysis, hypnagogic hallucinations, cultural influences, and the power of suggestion. While the possibility of extraterrestrial contact remains a fascinating and scientifically valid area of inquiry, a rigorous and unbiased analysis necessitates considering the limitations of relying solely on subjective accounts. By acknowledging the influence of these complex factors, we can move beyond simplistic explanations and embark on a more comprehensive exploration of the fascinating, yet enigmatic, world of reported alien abductions, recognizing the potential for insights that extend far beyond the extraterrestrial question. The enduring mystery of these reports encourages a multidisciplinary approach, combining the rigor of scientific investigation with the nuanced understanding provided by fields such as psychology, sociology, and anthropology, opening up avenues of research that offer the potential to further our understanding of the human mind and the myriad ways we perceive and interpret reality. The challenge lies in disentangling the subjective experiences from potential external influences, always acknowledging the complexities and limitations of human memory and perception.

The inherent difficulty in scientifically evaluating reported alien abduction accounts stems from the profound lack of verifiable physical evidence. Unlike other scientific investigations, where tangible data like fossils, astronomical observations, or chemical compounds provide a basis for analysis, abduction accounts rely almost entirely on eyewitness testimony. This immediately presents a significant hurdle. Human memory is notoriously fallible, susceptible to suggestion, distortion, and the influence of various psychological and social factors. The events described are often reported as occurring under conditions of intense stress and fear, circumstances known to significantly impair memory accuracy and recall. Furthermore, the elapsed time between the alleged event and the reporting of it introduces further opportunities for distortion and embellishment, influenced by discussions with others, media exposure, and the individual's own evolving understanding of the experience.

The problem of subjective reporting is compounded by the lack of standardized methodologies for collecting data on alleged abduction accounts. Many accounts are gathered through informal interviews, often conducted long after the reported event, leaving significant room for bias and inconsistencies. The absence of a structured, controlled data collection process makes it exceptionally challenging to isolate genuine phenomena from the effects of suggestion, cultural influences, and individual biases. The very act of asking someone about a potentially traumatic or extraordinary experience can lead to the recall of events that are either misremembered, confabulated, or amplified in response to the interview context. This is not to say that individuals are intentionally lying; rather, the process of recalling and narrating such experiences is inherently susceptible to distortion. It's a psychological process, not a neutral recording of facts.

This lack of standardized data collection methods makes comparisons between accounts extremely difficult. Even when accounts appear similar at a superficial level, the underlying details and contextual factors often differ significantly. Without careful control over data gathering, it's impossible to determine whether similarities reflect a shared experience or are simply the result of convergent narrative structures, influenced by media portrayals and cultural expectations. The absence of a standardized interview protocol, for example, prevents a meaningful assessment of the consistency and reliability of the information reported. Different interviewers might employ different techniques, inadvertently shaping the respondent's narrative in subtle but significant ways.

Furthermore, the absence of corroborating physical evidence poses a considerable challenge to scientific validation. In many accounts, advanced technologies or medical procedures are described. However, there is no independent verification of the existence of such technologies or any physical trace of their use on the individuals reporting these experiences. The lack of physical evidence, such as alien artifacts, biological samples, or traces of unusual energy signatures at alleged abduction sites, undermines the possibility of applying standard scientific methodologies to verify the accounts. The absence of this objective corroboration significantly weakens the claims, forcing us to rely entirely on subjective experience which, as previously discussed, is inherently prone to error and bias.

Another significant limitation stems from the difficulty in objectively assessing the psychological state of the individuals reporting abductions. While some individuals report experiencing significant psychological distress following the alleged event, it remains challenging to conclusively determine whether this distress is a direct result of an extraterrestrial encounter or is instead a product of other factors, such as sleep deprivation, pre-existing mental health conditions, or the stress associated with recounting a highly unusual and potentially stigmatizing experience. Establishing a clear causal link between the reported abduction and subsequent psychological symptoms requires rigorous psychological evaluation, accounting for various confounding factors. The absence of such systematic evaluations makes it difficult to definitively attribute observed psychological changes to the reported experience, rather than to other potential influences.

One approach to scientifically evaluating the reported accounts might involve a deeper exploration of sleep paralysis and other parasomnias. As previously mentioned, sleep paralysis, a common physiological experience, shares several characteristics with the reported events, including feelings of paralysis, hallucinations, and a sense of overwhelming dread. A detailed comparison of the subjective experiences during sleep paralysis with those described in alien abduction accounts could reveal commonalities and potential overlaps, offering alternative explanations for some aspects of the reported encounters. This requires a thorough investigation into the incidence of sleep paralysis within the populations reporting abductions, comparing these rates to control groups without reported experiences. Furthermore, analyzing the content of hallucinations during sleep paralysis could reveal potential parallels with the descriptions of abductors and abduction procedures.

Another avenue of investigation could involve the application of cognitive psychology principles to understand the processes of memory construction and recall. Studying the effect of stress, fear, and suggestion on memory reliability could shed light on the potential for distortion and confabulation in the abduction narratives. For instance, exploring the role of leading questions during interviews, the influence of media portrayals, and the impact of social pressure could provide valuable insights into the processes that contribute to the formation and evolution of these memories. This approach requires rigorous experimental designs, employing techniques such as cognitive interviews and memory testing to examine the accuracy and consistency of eyewitness accounts.

Furthermore, anthropological and sociological perspectives offer valuable frameworks for understanding the cultural context within which these accounts emerge. Analyzing the distribution of reports across different cultures, examining the prevalence of related beliefs and folklore, and considering the impact of media representations could provide insights into the social and cultural factors that shape the interpretation and reporting of these experiences. A comparative study across different cultural groups could reveal patterns and variations in the narratives, helping to differentiate experiences rooted in genuine events from those that reflect culturally shaped expectations and beliefs. This interdisciplinary approach is essential to move beyond simplistic explanations and gain a deeper understanding of the complex interplay between individual experiences and broader cultural narratives.

Finally, while the lack of physical evidence remains a substantial impediment to scientific validation, the ongoing search for extraterrestrial life is continually providing new insights and methodologies. The development of advanced technologies in fields such as astronomy, astrobiology, and biosignature detection could lead to future discoveries that may shed light on the possibility of extraterrestrial intelligence and their potential interaction with our planet. Although current evidence from abduction accounts remains insufficient for scientific verification, the exploration of the cosmos continues, and future discoveries may provide a new context for interpreting these enigmatic accounts. The field of astrobiology is not dismissing the possibility of alien contact; rather, it's emphasizing the importance of rigorous scientific methodology and a careful examination of alternative explanations before reaching definitive conclusions. The mystery of the alleged abductions presents a challenge, but one that is not outside the scope of future scientific investigation, as new technologies and methods continually refine our ability to understand the universe and our place within it.

In conclusion, the scientific evaluation of reported alien abduction accounts remains a significant challenge due to the absence of verifiable physical evidence and the limitations of eyewitness testimony. The lack of standardized data collection methods, the fallibility of human memory, and the influence of various psychological, social, and cultural factors all contribute to the difficulty in assessing the validity of these accounts. However, employing interdisciplinary approaches integrating psychological, sociological, and anthropological perspectives, alongside advances in astrobiology and related fields, holds the potential to offer a deeper understanding of these perplexing accounts, even if the ultimate question of extraterrestrial visitation remains unresolved for the time being. The journey to unraveling this mystery necessitates continuous investigation, open-mindedness, and rigorous scientific scrutiny.

The exploration of reported alien abduction accounts, while fraught with methodological challenges, offers a unique window into the complexities of human psychology. These accounts, regardless of their veracity concerning extraterrestrial involvement, reveal fundamental aspects of human experience, belief systems, and the human need for meaning-making. Analyzing the narratives, even without definitive proof of extraterrestrial contact, allows us to explore how humans process trauma, cope with the unknown, and construct narratives to explain extraordinary experiences.

One significant area of psychological interest lies in the examination of memory distortion and confabulation. The accounts often contain inconsistencies, details that shift over time, and elements that resemble common tropes found in science fiction and popular culture. These inconsistencies aren't necessarily indicative of deliberate deception; instead, they highlight the inherent limitations and malleability of human memory, particularly under conditions of extreme stress or trauma, as often reported in alleged abduction experiences. The act of recalling and recounting a highly unusual and potentially frightening event is not a simple replay of a recording; it's a reconstructive process influenced by various cognitive biases, social influences, and the individual's own expectations and beliefs. This process can lead to the creation of false memories, or the distortion of existing ones, to fit a coherent and understandable narrative.

Cognitive psychology offers valuable tools for dissecting these narrative processes. Studies on eyewitness testimony have shown that even seemingly minor details in questioning can significantly influence a person's recollection. Similarly, post-event information—discussions with others, media portrayals, or even the individual's own rumination—can subtly yet profoundly alter the narrative over time. In the context of alien abduction reports, the influence of cultural narratives surrounding extraterrestrial encounters is particularly potent. Repeated exposure to similar stories in movies, television, and literature can shape expectations and influence the way individuals interpret and recount their experiences. This doesn't necessarily imply fabrication, but rather illustrates the complex interplay between experience, memory, and cultural context in shaping personal narratives.

The phenomenon of sleep paralysis provides a compelling parallel for some elements of reported alien abduction accounts. Sleep paralysis, a relatively common experience, involves a temporary inability to move or speak while falling asleep or waking up, often accompanied by hallucinations. These hallucinations frequently involve a sense of pressure, an overwhelming feeling of dread, and even visual perceptions of shadowy figures or beings. The similarities between these hallucinations and descriptions of alien encounters in abduction reports are striking, leading some researchers to suggest that some, perhaps many, alleged abductions might be misinterpretations of sleep paralysis experiences, amplified and distorted through memory and cultural influences. A rigorous comparison of sleep paralysis experiences with reported abduction accounts, controlling for potential biases, is a crucial avenue for further research. This involves not only comparing subjective descriptions but also exploring underlying physiological and neurological factors.

Furthermore, the study of abduction accounts allows us to explore the psychology of belief and the human need for meaning. Many who report abduction experiences firmly believe in the truth of their accounts, often constructing elaborate and detailed narratives to explain their experiences. This unwavering belief highlights the potent influence of personal conviction and the human desire for coherent explanations, even in the face of contradictory evidence. The psychological factors that contribute to maintaining these beliefs, including emotional investment, social validation from similar belief communities, and the desire to make sense of a profoundly disturbing experience, are vital areas of inquiry. Understanding these psychological underpinnings can provide invaluable insights into the psychology of belief, its resilience in the face of skepticism, and the broader human need to find meaning and order in the world.

Beyond the individual level, alien abduction reports offer a unique lens through which to examine broader societal issues. These accounts often reveal deep-seated anxieties about technological advancements, societal control, and the unknown. The abductors, in the narratives, are frequently depicted as possessing advanced technology and exerting a degree of control over the abductees. This portrayal could be interpreted as a reflection of societal anxieties related to technological advancements, surveillance, and the potential loss of individual autonomy. Furthermore, the themes of bodily violation and experimentation found in many abduction narratives reflect anxieties surrounding the loss of bodily integrity and control over one's own physical self. Analyzing these narratives through a sociological perspective can offer insights into the anxieties and fears that shape our collective understanding of the world and our place within it.

The prevalence of certain common themes across different abduction accounts, despite variations in specific details, also warrants attention. The repetition of similar abduction scenarios, technologies, and procedures suggests the possibility of shared cultural narratives influencing individual accounts. This emphasizes the power of media portrayals and shared beliefs in shaping perceptions and interpretations of experiences. The consistency of certain narrative elements across widely separated geographical locations and cultural backgrounds raises questions about the extent to which individual experiences are genuinely independent or influenced by common cultural frameworks.

The study of abduction accounts also intersects with anthropological investigations of belief systems and folklore. Cross-cultural comparisons of these narratives might reveal common themes and variations that illuminate the broader context in which such experiences are interpreted and understood. By examining the role of pre-existing beliefs, cultural narratives, and indigenous folklore in shaping the interpretation of these encounters, we can gain a deeper appreciation for the complex interplay between individual experiences and broader cultural frameworks.

Finally, the continued investigation of alleged alien abduction reports, even in the absence of definitive proof of extraterrestrial contact, underscores the enduring fascination with the unknown and the human need for exploration and discovery. Whether the narratives are interpreted as genuine encounters, psychological phenomena, or products of cultural imagination, they serve as a compelling reminder of the profound mysteries surrounding human consciousness, the limits of human perception, and our place in the vast expanse of the cosmos. The rigorous scientific scrutiny these accounts demand, even if ultimately leading to the conclusion that they are not extraterrestrial in origin, offers valuable insights into the human condition and the powerful forces that shape our perceptions, beliefs, and understanding of reality. The exploration itself, therefore, is a significant contribution to understanding human psychology and society. It compels a deeper consideration of the factors that influence human memory, belief, and the ongoing human quest to comprehend the universe and our place within it.

| 6 |

The Implications of Contact

The possibility of contact with extraterrestrial life, while still firmly within the realm of speculation, opens a Pandora's Box of potential benefits that could reshape human civilization. The implications extend far beyond the simple confirmation of life beyond Earth; they represent a paradigm shift in our understanding of biology, physics, and our place in the cosmos. The potential advancements span across diverse scientific and technological fields, promising a future unimagined by previous generations.

One of the most immediate and potentially transformative benefits lies in the field of medicine. Imagine the possibility of encountering life forms with vastly different biological mechanisms and evolutionary pathways. Studying their physiology, genetics, and biochemistry could unlock revolutionary advancements in disease prevention, treatment, and even the extension of human lifespan. Extraterrestrial organisms might possess unique adaptations to environmental stresses – extreme temperatures, radiation, or lack of oxygen – offering insights into the development of novel therapeutic strategies for human ailments. For example, an alien organism's resilience to radiation could inspire the development of new radiation-resistant therapies for cancer treatment, or its ability to thrive in extreme cold might hold the key to treating hypothermia or other cold-related conditions. Their biological systems might harbor entirely new classes of antibiotics or antiviral compounds, providing solutions to the ever-growing problem of antibiotic resistance. The potential for discovering new pharmaceuticals or biomaterials with unique properties is virtually limitless, leading to advancements in regenerative medicine, organ transplantation, and the treatment of currently incurable diseases. Such advancements wouldn't just improve the quality of life; they could fundamentally alter our understanding of aging and disease itself.

Beyond medicine, the technological benefits of extraterrestrial contact are potentially staggering. The technologies employed by a sufficiently advanced civilization could far surpass our current capabilities, offering breakthroughs in fields like energy production, materials science, and transportation. Imagine accessing energy sources far exceeding the efficiency of our current methods, potentially solving the global energy crisis and drastically reducing our reliance on fossil fuels.
Extraterrestrial technologies might utilize principles of physics unknown to us, opening avenues for new forms of propulsion, faster-than-light travel, and even interdimensional travel, revolutionizing our ability to explore the cosmos. New materials with superior strength, durability, or conductivity could redefine engineering and construction, paving the way for more sustainable and efficient infrastructure. Even the communication technologies employed by an alien civilization could represent a leap forward, leading to breakthroughs in information

processing and data transmission that redefine our understanding of computing and communication. These technologies, once mastered, could reshape society, driving economic growth, technological innovation, and even fostering a new era of scientific discovery.

The societal impact of contact would be equally profound, extending into the realm of philosophy, ethics, and global cooperation. The simple act of discovering another intelligent species could reshape our understanding of our own place in the universe, shattering anthropocentric biases and promoting a more universal perspective. This could lead to a new era of scientific cooperation on a global scale, requiring a collaborative approach to unraveling the mysteries of extraterrestrial life and technology. Such collaborative efforts could potentially bridge existing geopolitical divides, uniting humanity in a shared endeavor to understand and interact with an alien civilization. Furthermore, the confrontation with an advanced civilization could force humanity to grapple with its own flaws and shortcomings, fostering a renewed commitment to environmental stewardship and social justice. The discovery of other intelligent life might provide a powerful impetus for global cooperation, pushing us towards a more unified and sustainable future.

However, the potential benefits must also be weighed against the potential risks. The technological advancement of an alien civilization might be so far ahead of our own that it could pose a threat to our existence, even unintentionally. Contact could lead to unforeseen ecological consequences, or even the introduction of hostile or dangerous biological agents. It is crucial that we develop strategies for responsible contact, considering potential dangers and establishing appropriate protocols for interaction. This requires a cautious and measured approach, prioritizing careful observation and analysis before any direct engagement. The development of international protocols and collaborations would be essential to mitigate potential risks and ensure the peaceful and beneficial interaction with other intelligent life forms. Scientific investigation and rigorous analysis should lead the way, ensuring that any form of contact is well-informed and guided by ethical considerations.

The discovery of extraterrestrial life, and the subsequent potential for contact, represents a profound turning point in human history. The potential benefits – advancements in medicine, technology, and societal understanding – are vast and transformative. However, the potential risks also necessitate careful planning and responsible action. By acknowledging both the opportunities and challenges, humanity can prepare for a future where contact with extraterrestrial life could be not only possible, but also profoundly beneficial for our species and our planet. The journey to prepare for such an encounter necessitates rigorous scientific research, international collaboration, and a commitment to ethical considerations that will guide our interactions with the vast unknown that lies beyond Earth. The challenge lies not just in seeking out life beyond our planet, but in ensuring that our first contact is a moment of cooperation, understanding, and a shared journey toward a brighter future for all intelligent life in the universe. This requires a fundamental shift in our approach to science, technology, and global cooperation, fostering a new era of collaboration driven by a shared sense of wonder and a commitment to the peaceful exploration of the cosmos.

The prospect of encountering extraterrestrial intelligence, while brimming with potential benefits, also presents a range of significant risks. These risks aren't merely science fiction scenarios; they necessitate careful consideration and proactive mitigation strategies. The sheer unknown surrounding extraterrestrial biology, technology, and societal structures necessitates a cautious approach. Failure to adequately assess and address these risks could have catastrophic consequences for humanity.

One of the most immediate concerns is the potential for conflict. A technologically superior civilization might view humanity as a threat, a resource to be exploited, or simply an obstacle to their own goals. Even unintentional conflict is a possibility; a misunderstanding or misinterpretation of our actions could escalate into a disastrous confrontation. This scenario highlights the crucial need for developing sophisticated communication strategies and protocols – a universal language, if possible – to ensure clear and unambiguous interactions. The development of such protocols requires extensive interdisciplinary collaboration, drawing upon expertise from linguistics, anthropology, sociology, and political science, to anticipate potential misunderstandings and design methods for avoiding conflict. The inherent risks of miscommunication are amplified by the possibility of vastly different cultural norms and values, potentially leading to unintended offense or aggression. Consider the challenges humans face in communicating effectively across vastly different cultures on Earth; the challenges in interspecies communication would be exponentially greater.

Beyond direct conflict, the threat of exploitation presents a significant risk. A more advanced civilization might be interested in our planet's resources, our genetic material, or our technological capabilities. They could subtly manipulate our societies or overtly seize control of our resources, leaving humanity marginalized or enslaved. This scenario underscores the importance of carefully controlling the dissemination of information about our planet and its capabilities. Transparency could be a double-edged sword; while cooperation necessitates some sharing of information, it's crucial to carefully manage what we reveal to avoid unintended consequences. Strategic information management, involving global cooperation and careful consideration of all potential outcomes, would be paramount in protecting humanity's interests. Furthermore, international cooperation would be vital to present a unified front and avoid internal conflicts that might weaken our position.

The transmission of disease represents another significant threat. Extraterrestrial organisms may harbor pathogens against which our immune systems have no defense. Such pathogens could cause devastating pandemics, potentially wiping out entire populations or even driving humanity to extinction. Conversely, we could unknowingly transmit terrestrial diseases to extraterrestrial life, with potentially devastating consequences for them and potentially jeopardizing any possibility of future positive interactions. This underscores the absolute necessity of stringent quarantine protocols, both for any extraterrestrial samples brought back to Earth and for any human missions venturing into extraterrestrial environments. Extensive research into planetary protection protocols, building upon existing models used for preventing the contamination of other planets in our solar system, will be crucial to minimize the risk of accidental cross-contamination. This requires international cooperation and a commitment to rigorous scientific standards in all aspects of extraterrestrial exploration.

Beyond biological threats, we must also consider the possibility of technological risks. An advanced civilization might possess technologies that could be used to manipulate our environment, our minds, or our social structures in ways we cannot even imagine. Such technologies could destabilize global systems, causing widespread chaos or enabling oppressive control. The potential for technological subversion, whether through cyberattacks or other forms of interference, demands proactive development of robust defensive systems and strategies. This requires investments in cybersecurity, artificial intelligence, and other relevant fields, while fostering international cooperation to share threat intelligence and coordinate responses. The potential for unknown technologies to have unforeseen and catastrophic consequences highlights the necessity of cautious and incremental approaches to any form of contact.

The ethical implications of contact also pose serious challenges. How do we interact with a civilization that might possess vastly different values, morality, or even a concept of selfhood? How do we avoid imposing our own cultural norms and values on them, and how do we protect our own culture from potentially harmful influences? These questions demand careful ethical reflection and the development of internationally agreed-upon principles that guide our interactions. This will require considerable philosophical and ethical debate to establish a common framework for decision-making, taking into account diverse cultural perspectives and ensuring fairness and respect for all involved. A robust ethical framework will be essential to avoiding unintended consequences and fostering mutually respectful relationships.

The sheer unpredictability of contact with an extraterrestrial civilization underscores the necessity of preparing for a wide range of scenarios. Contingency planning is essential; we must develop robust strategies to respond to a variety of potential situations, from peaceful cooperation to outright conflict. This necessitates an integrated approach, combining scientific research, technological development, diplomatic strategy, and ethical considerations. The development of robust contingency plans needs to be a continuous process, involving regular review and updates based on new data and insights. Furthermore, public education and engagement are vital; preparing the global population for the possibility of contact will help to ensure a unified and informed response.

In conclusion, the potential risks associated with contact with extraterrestrial intelligence are significant and varied. They range from direct conflict and exploitation to the transmission of diseases and technological disruption. Addressing these risks requires a multifaceted approach, combining scientific research, international collaboration, ethical reflection, and robust contingency planning. Preparing for the possibility of contact is not just a scientific endeavor; it is a global undertaking that demands a unified and coordinated response, guided by a commitment to responsible exploration and the preservation of humanity's future. The journey towards potential contact necessitates not only the advancement of our scientific capabilities but also a profound reassessment of our values, ethics, and our place within the vastness of the cosmos. Only through a thoughtful, well-planned, and internationally coordinated strategy can we hope to navigate the potential risks and harness the benefits of this momentous event in human history.

The ethical dimensions of potential contact with extraterrestrial intelligence are profound and multifaceted, demanding careful consideration far beyond the immediate concerns of technological or biological risks. While the potential for conflict and disease warrants significant attention, the very act of encountering another intelligent species raises fundamental questions about our responsibilities, our values, and our place in the cosmos. These ethical considerations are not merely hypothetical; they require proactive engagement and the development of robust frameworks to guide our actions.

One of the most pressing ethical dilemmas concerns communication protocols. Assuming we establish contact, how do we communicate meaningfully with a civilization that may have a vastly different understanding of language, concepts, and even reality itself? The development of a universal language, or a system of communication that transcends linguistic barriers, is a monumental task, demanding collaboration between linguists, anthropologists, and experts in artificial intelligence. Furthermore, the content of our initial messages requires careful deliberation. Do we reveal our technological capabilities, our societal structures, or our history? What information is necessary for establishing peaceful communication, and what might inadvertently invite hostility or exploitation? The choices we make in crafting our first messages will have far-reaching consequences, impacting the trajectory of our relationship with this other civilization.

Beyond the initial contact, ethical considerations extend to resource sharing and exploitation. If an extraterrestrial civilization possesses superior technology or resources, how do we ensure a fair and equitable distribution of benefits? Should we share our planet's resources, even if it poses a risk to our own survival? Conversely, how do we protect our resources from exploitation by a potentially more powerful civilization? These questions raise complex issues of global governance and international cooperation. A unified, global strategy will be crucial to negotiate interactions that are both mutually beneficial and protective of humanity's interests. Failure to establish clear ethical principles and collaborative mechanisms could lead to exploitation and the marginalization of humanity on a cosmic scale.

The treatment of extraterrestrial beings is another critical ethical concern. If we encounter extraterrestrial life, what obligations do we have towards them? Should we adhere to principles of non-interference, respecting their sovereignty and autonomy? Or do we have a responsibility to assist them, perhaps even intervene in their development if we perceive them to be in danger? The answers to these questions demand a reassessment of our anthropocentric worldview and the development of ethical frameworks that extend beyond human-centric values. The ethical treatment of extraterrestrial life necessitates recognizing their inherent worth and dignity, regardless of their technological advancement or perceived level of intelligence. This may require grappling with philosophical questions about sentience, consciousness, and the very definition of life itself.

The potential for genetic manipulation presents further ethical complexities. Imagine encountering a civilization with vastly different biological structures or genetic makeup. The temptation to study their biology, potentially for scientific advancement or technological enhancement, might be strong. However, such actions could have unforeseen consequences, jeopardizing their survival or violating their inherent right to exist free from external manipulation. Respecting the integrity of extraterrestrial life forms necessitates establishing strict ethical guidelines for scientific research and exploration, guaranteeing that all interactions are conducted with the utmost care and consideration for their well-being. This requires a shift in our scientific paradigm, moving beyond a purely exploitative approach to one that prioritizes understanding and respect.

Cultural exchange also presents significant ethical challenges. Contact with an extraterrestrial civilization would inevitably involve a cultural exchange, presenting opportunities for learning and growth but also potential risks of cultural contamination or dominance. How do we ensure a mutually respectful exchange of ideas and knowledge, avoiding the imposition of our own cultural norms and values? How do we protect our own cultures from potentially harmful influences? These questions demand a thorough examination of our own biases and assumptions, fostering intercultural understanding and developing strategies for navigating intercultural interactions with grace and sensitivity. The process of cultural exchange requires careful diplomacy and a commitment to mutual learning and respect.

The issue of planetary protection extends beyond merely preventing the transmission of terrestrial diseases. It necessitates recognizing the potential impact of our actions on extraterrestrial ecosystems. Any exploration or interaction with extraterrestrial environments needs to be conducted in a way that minimizes harm to any existing life forms, whether microbial or more complex. This involves developing stringent protocols for sanitation, waste disposal, and the avoidance of any actions that might disrupt or damage delicate extraterrestrial ecosystems. Such protocols need to be based on rigorous scientific understanding and a commitment to environmental stewardship on a cosmic scale. Failure to adopt such measures could lead to irreversible ecological damage, jeopardizing the very possibility of understanding the life we may encounter.

Finally, the ethical implications of contact are intimately tied to questions of global governance and international cooperation. The potential benefits and risks of contact are shared by all of humanity; therefore, the decisions regarding how to interact with extraterrestrial intelligence should not rest with a single nation or group but should involve a global consensus. Establishing an international body to manage the process of contact, ensuring transparency, accountability, and the equitable distribution of resources and knowledge, will be crucial to navigating the challenges and harnessing the opportunities presented by this momentous event. This requires building trust and fostering a spirit of cooperation among nations, transcending national interests for the benefit of all humankind.

In conclusion, the ethical implications of contact with extraterrestrial intelligence are vast and complex, touching upon fundamental questions about our responsibilities, values, and place in the universe. Addressing these challenges requires a proactive and multifaceted approach, involving not only scientific research but also philosophical inquiry, ethical reflection, and robust global governance. The development of ethical frameworks for interstellar interaction is not simply a philosophical exercise; it is a necessity for ensuring the survival and flourishing of humanity and the respect of any extraterrestrial life we might encounter. Only through careful consideration, thoughtful planning, and unwavering commitment to ethical principles can we hope to navigate the profound challenges and harness the incredible potential of contact with other intelligent life in the cosmos. The journey towards potential contact necessitates not only the advancement of our scientific capabilities but also a profound reassessment of our values, ethics, and our place within the vastness of the cosmos.

The prospect of contact with extraterrestrial intelligence necessitates a paradigm shift in global governance. The sheer scale and potential ramifications of such an event transcend national borders and individual interests, demanding a unified, internationally coordinated response. A haphazard, piecemeal approach, characterized by nationalistic rivalries and a lack of global consensus, would almost certainly lead to chaos, misunderstanding, and potentially catastrophic consequences. Therefore, the establishment of a robust international framework for managing extraterrestrial contact is not merely desirable;

it is absolutely essential for the survival and well-being of humanity.

Several models for international cooperation present themselves. One approach might involve strengthening existing international organizations like the United Nations. The UN's structure, while imperfect, offers a pre-existing framework for global dialogue and cooperation. Assigning the task of managing potential extraterrestrial contact to a newly empowered committee within the UN, perhaps a specialized agency akin to the World Health Organization or the International Atomic Energy Agency, could leverage existing diplomatic channels and expertise. This body could serve as a central hub for information gathering, scientific research coordination, and the formulation of international protocols for communication and interaction.

However, simply expanding the UN's remit might prove insufficient. The complexities and unprecedented nature of extraterrestrial contact necessitate a more agile, specialized structure. This could take the form of a new international body, possibly independent of the UN, but working in close collaboration with it. This new organization could be tasked with specific responsibilities, including:

Protocol Development: Creating and disseminating standardized protocols for communication, data sharing, and interaction with extraterrestrial civilizations. This would involve the collaboration of scientists, linguists, anthropologists, and ethicists from across the globe, working together to develop a common language or communication system and a set of ethical guidelines. The challenge would lie in navigating differing cultural interpretations and value systems, establishing consensus on acceptable methods of engagement and building a collective understanding of the implications of contact.

Resource Management: Developing strategies for the equitable distribution of resources and benefits derived from extraterrestrial contact. This necessitates a sophisticated global mechanism to prevent exploitation, ensure fair access to resources, and safeguard humanity's interests, balancing potential gains with the need for responsible management and sustainability. Negotiating agreements on resource allocation in the face of vastly different capabilities or needs would require unprecedented levels of international trust and cooperation.

Risk Assessment and Mitigation: Conducting ongoing assessments of potential risks associated with extraterrestrial contact, including biological hazards, technological threats, and socio-cultural impacts. This would involve a global network of scientists, experts, and intelligence agencies working in concert to identify and mitigate risks, developing emergency plans to manage any unforeseen events.

Public Education and Engagement: Developing and implementing programs for public education and engagement on the issue of extraterrestrial contact. Transparency and open communication with the public are vital in fostering understanding, addressing concerns, and promoting global cooperation in the face of such a momentous event. Public education campaigns could address various aspects of the subject, fostering both scientific literacy and ethical sensitivity concerning the implications of contact.

The success of any international governance structure depends heavily on the principles upon which it is founded. Transparency, accountability, and inclusivity are paramount. All nations, regardless of their political systems or economic power, must have a voice in the decision-making process. A global, participatory approach is necessary to ensure equitable representation and avoid the domination of any single nation or group. The rules of engagement should be based on universally accepted ethical principles, incorporating the values of respect, non-interference, and mutual benefit. Furthermore, robust mechanisms for dispute resolution and conflict management need to be embedded within the structure, ensuring peaceful resolution of disagreements.

The challenges in achieving effective international cooperation are significant. National self-interest, political rivalries, and differing cultural perspectives can impede progress. Building trust among nations, particularly those with conflicting geopolitical agendas, will be crucial. Transparency in data sharing and decision-making processes is fundamental in fostering trust and preventing the emergence of secretive, potentially harmful agendas. Collaborative scientific research is a key pillar in building this trust, showcasing shared goals and the imperative for a unified approach. The creation of a global scientific consortium, dedicated to research on extraterrestrial phenomena, could serve as a powerful symbol of unity and shared purpose.

The governance structure must also address the potential for internal conflicts. Disagreements over resource allocation, communication strategies, or risk mitigation measures could easily emerge. The framework must incorporate effective mechanisms for resolving these disputes, perhaps through arbitration or mediation processes. It is crucial to establish a clear decision-making hierarchy, while simultaneously ensuring the involvement of a broad range of experts and stakeholders, to avoid being overly centralized and potentially autocratic. Furthermore, the framework must accommodate the dynamic nature of the situation; it should be adaptable to new information and unforeseen developments.

Beyond formal governance structures, informal networks and collaborations will also play a vital role. Scientific exchanges, diplomatic initiatives, and public forums can all contribute to fostering a global sense of shared responsibility. The involvement of civil society organizations, including scientific societies, academic institutions, and ethical think tanks, will be invaluable in shaping international policy and ensuring that the perspectives of various stakeholders are considered. Encouraging public discourse and promoting ethical debate will help build societal consensus and ensure that humanity's response to extraterrestrial contact is guided by shared values and responsible decision-making.

In conclusion, the implications of extraterrestrial contact demand not only scientific preparedness but also a radical transformation in global governance. Establishing a robust, internationally cooperative structure will be a monumental task, demanding extraordinary levels of cooperation, trust, and shared vision. However, the stakes are too high to ignore. Failure to establish effective international governance mechanisms in the face of extraterrestrial contact could lead to conflict, exploitation, and catastrophic consequences. A successful approach will require an unprecedented level of global cooperation, a shared commitment to ethical principles, and a willingness to overcome national self-interest for the sake of humanity's collective future. The creation of such a framework represents one of the most critical challenges—and opportunities—facing humanity in the 21st century.

Preparing for the unknown is not simply a matter of reacting to events as they unfold; it demands a proactive and comprehensive strategy. The potential implications of contact with extraterrestrial intelligence are so profound and far-reaching that a reactive approach would be dangerously inadequate. Instead, we must develop a robust framework for preparedness that anticipates a range of scenarios and equips humanity to navigate this unprecedented moment in our history responsibly. This necessitates a multi-faceted approach, encompassing scientific research, international collaboration, and ethical considerations.

One of the most crucial aspects of preparation involves bolstering our scientific understanding of the universe and the potential for life beyond Earth. This isn't solely about searching for extraterrestrial signals; it's about broadening our knowledge across various disciplines. Astrobiology, the study of life's origins and evolution in the universe, will play a vital role, providing a framework for understanding the possible characteristics of extraterrestrial life, their potential technological capabilities, and their potential impact on Earth's ecosystems. Advances in fields like genomics, exoplanet research, and advanced materials science are crucial for understanding potential biological hazards, interpreting potential extraterrestrial communication, and developing technologies that might be necessary for interaction.

A critical component of scientific preparedness lies in the development of advanced communication technologies. The challenge of communicating across vast interstellar distances requires innovative approaches. We need to develop sophisticated methods for encoding and transmitting messages that are unambiguous and easily understood, even by a civilization vastly different from our own. This necessitates not just technological breakthroughs but also a deeper understanding of communication theory, linguistics, and semiotics—the study of signs and symbols. Developing a universal language, or at least a framework for interspecies communication, will require the collaborative efforts of linguists, anthropologists, and computer scientists. Consider the complexities: even communication between different human cultures presents significant challenges; imagine the difficulty of communicating with a species that may have evolved under completely different environmental and evolutionary pressures.

Beyond communication, we must also anticipate the potential risks associated with extraterrestrial contact. This includes the possibility of encountering hostile or exploitative civilizations. But the dangers are not limited to aggression. Accidental exposure to unknown pathogens, technological disruptions, or unforeseen ecological consequences pose significant threats that demand careful consideration and the development of robust mitigation strategies. This requires collaboration between scientists, epidemiologists, and cybersecurity experts, along with the establishment of international monitoring and early warning systems. We might need to prepare for potential quarantine procedures, develop advanced detection systems for extraterrestrial pathogens, and establish protocols for responding to technological threats, such as malware or advanced weaponry.

International collaboration is paramount in addressing the challenges of extraterrestrial contact. A unified global response is not merely desirable, it's essential for humanity's survival. A piecemeal, nation-state-centric approach would be chaotic, fraught with mistrust, and potentially disastrous. The establishment of a robust international framework, perhaps under the auspices of the United Nations, or a newly created independent body, is essential. This framework would need to incorporate several key elements:

Firstly, it must establish standardized protocols for communication and interaction with extraterrestrial civilizations. These protocols would need to be developed collaboratively, ensuring the participation of diverse voices from across the globe, considering various cultural, ethical, and philosophical perspectives. This would require navigating potential conflicts in values and priorities, reaching consensus on ethical guidelines for first contact, and developing a system for ensuring that interactions are conducted in a manner that is respectful, non-exploitative, and mutually beneficial.

Secondly, the framework must address the potential for conflict and competition arising from the discovery of extraterrestrial resources. A transparent and equitable system for resource management and allocation is crucial, preventing the emergence of exploitative practices or conflict between nations. This requires the development of international legal frameworks that govern the extraction and use of resources obtained from extraterrestrial sources, ensuring fair access and preventing the concentration of resources in the hands of a few powerful entities.

Thirdly, the framework must incorporate mechanisms for risk assessment and mitigation. This necessitates the establishment of a global network of scientists, experts, and intelligence agencies, cooperating to monitor potential risks, develop emergency response protocols, and ensure a coordinated approach to mitigating any unforeseen consequences of contact. The potential risks are broad-ranging, from the introduction of new diseases to unexpected technological disruptions or even existential threats. Preparing for such contingencies necessitates rigorous modeling, scenario planning, and the development of effective emergency response procedures.

Furthermore, robust mechanisms for public education and engagement are essential. Open communication with the public builds trust, counters misinformation, and promotes global cooperation. It's crucial that people understand the scientific basis of the search for extraterrestrial life, the potential implications of contact, and the ethical considerations involved. Comprehensive public education programs can help foster a sense of shared responsibility and ensure that humanity approaches this unprecedented challenge in a thoughtful and collaborative manner.

The ethical considerations surrounding extraterrestrial contact are equally crucial. We must develop a clear set of ethical guidelines to govern our interactions with any extraterrestrial civilization. These guidelines should be based on universally accepted principles of respect, non-interference, and mutual benefit. They must also address potential conflicts of interest, ensuring that any interactions are conducted in a manner that respects the autonomy and rights of any extraterrestrial life forms encountered. Consider the complexities of potential cultural misunderstandings, technological disparities, and the ethical dilemmas arising from the possibility of encountering a species with vastly different values and social structures.

Preparing for the unknown requires not just scientific and technological advances but also a fundamental shift in our approach to international cooperation and ethical decision-making. It's a challenge that necessitates a global effort, underpinned by a commitment to collaboration, transparency, and a shared sense of responsibility for the future of humanity. Failure to prepare effectively could have catastrophic consequences. Success will hinge on our ability to overcome national self-interest, foster trust among nations, and develop a collective response guided by ethical principles and a shared vision for humanity's future. This is not simply about preparing for the possibility of contact; it is about shaping humanity's destiny in the face of the universe's vastness and the profound unknown that lies beyond our planet. The journey into this unknown requires careful planning, collaboration on a global scale, and a commitment to ethical decision-making—a journey that will define humanity's future for millennia to come.

| 7 |

The Fermi Paradox

The Fermi Paradox, named after the physicist Enrico Fermi, encapsulates a profound contradiction at the heart of our search for extraterrestrial life. It hinges on a simple, yet unsettling question: given the vastness of the universe, the sheer number of stars, and the billions of years it has existed, why haven't we encountered any evidence of other intelligent civilizations? The paradox arises from the seemingly high probability of extraterrestrial life, based on our understanding of cosmology and biology, clashing with the conspicuous absence of any contact or detectable signs of advanced extraterrestrial intelligence (SETI).

One of the foundational premises of the Fermi Paradox is the sheer scale of the universe. Our Milky Way galaxy alone contains hundreds of billions of stars, many of which are likely to have planetary systems orbiting them. Extrapolating this to the observable universe, with its billions of galaxies, each containing billions of stars, leads to an astronomically large number of potential habitable worlds. The probability of life arising on at least one of these planets, given the right conditions, seems statistically significant, even if the probability of life arising on *any* given planet is extremely low. The sheer number of chances significantly increases the likelihood of life emerging somewhere else.

Further bolstering this probability is our growing understanding of extremophiles—organisms that thrive in environments previously considered uninhabitable. The discovery of life in extreme conditions on Earth, such as deep-sea hydrothermal vents, acidic hot springs, and even within nuclear reactors, suggests that life may be far more resilient and adaptable than we previously thought. This expands the potential range of habitable environments within and beyond our solar system, further increasing the chances of finding extraterrestrial life. If life can flourish in such harsh conditions on Earth, it stands to reason that it could also exist in equally extreme, yet potentially habitable, environments on other planets. The definition of "habitable" itself is constantly evolving as our understanding of extremophiles expands.

Adding another layer to this probability calculation is the vast timescale involved. The universe is approximately 13.8 billion years old. Earth, and life on it, are relatively young by cosmic standards. If life evolved on Earth relatively quickly after the planet formed, then it's reasonable to assume that life could have evolved on other planets billions of years before Earth even existed. This immense timescale allows for the possibility of civilizations emerging, advancing technologically, and even potentially perishing long before humanity even appeared on the scene. Such civilizations could have reached incredibly advanced technological levels, perhaps even beyond our current comprehension, and left behind evidence that we haven't yet discovered, or which we are simply incapable of detecting with our current technology.

The paradox deepens when we consider the potential for interstellar travel. While the distances involved are immense, presenting significant technological challenges, the sheer timescale of the universe makes interstellar travel, theoretically, plausible. Even if a civilization's technology only allowed for relatively slow interstellar travel, over billions of years, they could have colonized a significant portion of the galaxy. This leads to the question: if interstellar travel is possible, and if numerous advanced civilizations have existed, why haven't they made contact with us, or at least left behind detectable signs of their presence? This is the core of the Fermi Paradox's unsettling nature.

The lack of evidence for extraterrestrial civilizations, despite the seemingly high probability of their existence, presents a significant challenge to our assumptions. Several hypotheses have been proposed to resolve or at least mitigate this apparent contradiction. One possibility is the "Great Filter" hypothesis, which suggests that there is a significant hurdle or obstacle that prevents the emergence or survival of intelligent life. This filter could lie in the past, meaning that the formation of life, the development of intelligence, or the transition to a technological civilization are exceptionally rare events. Alternatively, the filter could lie in the future, perhaps representing a catastrophic event that almost invariably wipes out advanced civilizations before they can reach interstellar travel capabilities or leave detectable signals. This catastrophic event could range from self-destruction through nuclear war or environmental collapse to unforeseen technological accidents or even natural disasters on a galactic scale.

Another hypothesis suggests that interstellar travel or communication might be far more difficult or costly than we currently imagine. The distances involved, the energy requirements, and the technological challenges are considerable. The challenges of interstellar communication are equally significant. Developing technologies capable of transmitting signals across vast interstellar distances and deciphering potentially alien communications present enormous hurdles. Furthermore, the potential for misinterpreting signals, or failing to recognize them as originating from an intelligent source, should not be discounted. The universe may be filled with signals, but we lack the capacity to understand them.

The possibility that advanced civilizations may choose not to contact us or reveal their presence is also a significant factor. This could stem from various motivations, ranging from self-preservation and a desire to avoid interference with less-developed civilizations to a lack of interest in contact. This "zoo hypothesis" suggests that advanced civilizations might observe Earth from a distance, choosing not to interfere.

Furthermore, the "Rare Earth Hypothesis" posits that the conditions necessary for the emergence of complex life, particularly intelligent life, are exceptionally rare and specific. While the universe may be teeming with life in simpler forms, the evolution of complex organisms, possessing intelligence and technology, might require a unique confluence of factors, making Earth a highly unusual case.

It's crucial to acknowledge the limitations of our current knowledge. Our search for extraterrestrial life is still in its infancy. Our technology for detecting extraterrestrial signals is constantly evolving, and we are only beginning to understand the vast diversity of life that might exist elsewhere in the universe. The possibility of life forms vastly different from anything we can currently imagine is very real and cannot be ruled out.

The Fermi Paradox remains a compelling puzzle, stimulating ongoing research and discussion in astrobiology, cosmology, and SETI. It forces us to reconsider our assumptions about the prevalence of life in the universe and encourages a more rigorous examination of the possibilities and limitations of our current scientific understanding. While there's no definitive answer, the paradox serves as a crucial reminder of the vastness of the unknown and the profound implications of discovering, or not discovering, extraterrestrial life. The ongoing search for extraterrestrial intelligence is not merely a scientific endeavor; it is a profound exploration of humanity's place within the cosmos and a reflection on our own future. The resolution, or lack thereof, of the Fermi Paradox could fundamentally reshape our understanding of the universe and our place within it.

The Fermi Paradox, with its unsettling silence in the face of seemingly overwhelming probabilities, has spurred a wealth of proposed resolutions. These attempts to reconcile the vastness of the universe teeming with potential life and the conspicuous absence of any confirmed extraterrestrial contact represent a fascinating blend of scientific speculation, philosophical pondering, and technological considerations. Let's delve into some of the most prominent and credible explanations offered to date.

One of the most enduring and thought-provoking solutions is the "Great Filter" hypothesis. This concept postulates that there exists a significant hurdle, a kind of cosmic bottleneck, that prevents the emergence or survival of advanced, technologically capable civilizations. This filter could lie in the past, meaning that the very emergence of life, the development of intelligence, or the transition to a technological society are exceptionally improbable events. Consider the sheer number of steps required for life to arise: the formation of suitable planets, the emergence of self-replicating molecules, the development of cellular structures, the evolution of multicellular organisms, the eventual appearance of intelligence, and finally, the development of a technological civilization capable of interstellar communication or travel. Each step represents a potential filter, a point of failure where the chain of events could be broken. If any one of these steps is exceptionally improbable, it could explain the absence of widespread extraterrestrial civilizations. A low probability at any single stage, multiplied across the numerous stages, would dramatically reduce the overall probability of a civilization reaching our level of development, let alone surpassing it.

Alternatively, the Great Filter could lie in the *future*. This more pessimistic perspective suggests that a catastrophic event, almost inevitable for technologically advanced civilizations, prevents them from surviving long enough to spread throughout the galaxy or leave detectable signals. This catastrophic event could take many forms. Self-destruction through nuclear war, ecological collapse driven by unsustainable resource consumption, the rise of uncontrollably destructive artificial intelligence, or unforeseen technological accidents are all possibilities. Perhaps there's a natural cosmic hazard, such as a particularly devastating gamma-ray burst or asteroid impact, that regularly decimates advanced civilizations before they can establish interstellar empires. Under this scenario, we might be witnessing a period of relative calm before the inevitable arrival of our own Great Filter. The unsettling implication is that we might be exceptionally naive to believe we'll somehow avoid this fate. This perspective, while bleak, underscores the urgency of mitigating risks in our own technological development and emphasizing sustainability.

The sheer difficulty of interstellar travel and communication provides another potential explanation for the Fermi Paradox. While the vastness of space is a considerable hurdle, the technological challenges are even more daunting. The distances involved are almost incomprehensibly immense, requiring either incredibly advanced propulsion systems or impossibly long travel times. Even with advanced technology, maintaining a functional spacecraft and sustaining a crew on such a voyage would pose immense difficulties. The resources required to launch and maintain such endeavors are potentially astronomical. The energy requirements alone could be far beyond the capacity of even the most advanced civilizations, acting as a practical limit to expansion. Furthermore, the energy required for interstellar communication, capable of being reliably detected across light-years, is also substantial. Transmitting and receiving such signals necessitates advanced technology and an understanding of the potential communication methods and challenges of interstellar propagation.

The "zoo hypothesis," a more speculative suggestion, proposes that advanced civilizations may be aware of Earth and our inhabitants but have chosen not to interfere. This idea suggests a kind of cosmic "hands-off" policy, perhaps stemming from a sense of preservation, an ethical imperative not to disturb less-developed civilizations, or simply a lack of interest in contact. The analogy of observing animals in a zoo illustrates the concept. These advanced extraterrestrials might be observing humanity's progress, waiting to see whether we meet certain criteria before engaging. Such a "zoo" scenario could also be less deliberate and more of a passive observation. Perhaps the necessary technology or the willingness to engage are simply not present. This idea lends itself to considerable debate and leaves plenty of room for speculation. One could even argue that their decision to not interfere is itself a form of interference. The act of observation alone could subtly alter humanity's course of development.

The "Rare Earth Hypothesis" suggests a different approach to the paradox. This hypothesis posits that the specific confluence of factors necessary for the emergence of complex life, particularly intelligent life, is exceedingly rare. While life might be relatively common throughout the universe in simpler forms, the evolution of organisms possessing intelligence and technological capabilities may demand a unique combination of planetary conditions, stellar characteristics, and geological events. Earth's characteristics, including its size, orbit, composition, plate tectonics, and the presence of a large moon, could all be considered essential elements in the emergence of intelligent life. Under this hypothesis, Earth may be an anomaly, a peculiar confluence of factors leading to a remarkable outcome, rather than a typical example of planetary evolution. The implications of this are that despite the vastness of the universe, planets with the necessary conditions for complex life may be few and far between.

Beyond these more established hypotheses, other possibilities exist. Our understanding of life itself might be too limited. We might be searching for life that conforms to our own understanding of biology, neglecting the possibility of entirely different biochemistries and life forms. The universe's age is also a factor; perhaps the rise and fall of civilizations are so rapid relative to cosmic timescales that the "overlap" of multiple civilizations is exceedingly rare. This could explain why we haven't detected signals – the window of opportunity for simultaneous existence might be incredibly short.

The limitations of our current technology must also be acknowledged. Our ability to detect extraterrestrial signals is still relatively primitive, our searches are limited in scope, and our understanding of what constitutes a "detectable signal" is still developing. It's possible that advanced civilizations communicate or signal in ways we haven't even begun to imagine, rendering our current search methods ineffective. We may also be interpreting the data incorrectly, overlooking signals that actually exist but are misinterpreted as naturally occurring phenomena.

The Fermi Paradox remains a compelling and unsolved puzzle. It serves as a powerful impetus for continued research in astrobiology and the search for extraterrestrial intelligence. It encourages a more rigorous scrutiny of our assumptions about the prevalence of life and challenges us to expand our thinking beyond our current limitations. The resolution of this paradox, whenever it might occur, will undoubtedly have a profound impact on our understanding of our place in the universe and our future as a species. The silence of the cosmos continues to inspire both wonder and concern, fueling the ongoing exploration of this enduring cosmic enigma.

Evaluating the plausibility of each resolution requires a careful weighing of evidence and a frank acknowledgment of the inherent uncertainties. Let's examine each proposed solution in turn, analyzing its strengths and weaknesses within the context of our current scientific understanding.

The **Great Filter** hypothesis, while elegant in its simplicity, suffers from a significant drawback: its inherent untestability. While we can speculate on potential filters, whether past or future, proving which, if any, is the actual bottleneck remains beyond our current capabilities. The argument that the emergence of life itself is an incredibly improbable event faces the counterargument of the sheer abundance of life on Earth, demonstrating that at least once, the process occurred. However, this doesn't negate the possibility that this was a singular event, a cosmic lottery win with incredibly long odds. The future filter scenario, while alarming, offers little in the way of concrete evidence. We can identify potential threats, such as nuclear war or artificial intelligence gone rogue, but predicting their likelihood with any certainty is impossible. Even natural cosmic events, such as gamma-ray bursts, while undeniably destructive, occur with a frequency that doesn't necessarily explain the complete absence of detectable signals from other civilizations.

The difficulty of **interstellar travel and communication** presents a more concrete challenge. While the vast distances involved are undeniable, technological advancements may render some of these challenges less insurmountable than they appear today. Our understanding of propulsion systems is still in its relative infancy. The development of faster-than-light travel, while currently considered science fiction, cannot be entirely ruled out. Furthermore, alternative communication methods might exist beyond our current detection capabilities. The energy requirements, while significant, are not necessarily insurmountable for a sufficiently advanced civilization. While energy demands pose a challenge, the limitations might be more of a practical restriction than an absolute impossibility for a civilization with sufficiently advanced energy harvesting and management capabilities. It's important to acknowledge that we are operating with a very narrow understanding of what constitutes "advanced technology." An alien civilization might possess technologies we can't even conceive of.

The **zoo hypothesis**, intriguing as it is, borders on pure speculation. There is no concrete evidence to suggest that advanced civilizations are deliberately avoiding contact, nor any strong theoretical basis for predicting such behavior. While it's certainly possible, the lack of empirical data makes it difficult to assess its plausibility. Moreover, the concept of a "zoo" presupposes motivations and ethical frameworks that might not apply to alien intelligences, vastly different from ourselves. The notion of a passively observant civilization is equally speculative, leaving a substantial room for philosophical arguments rather than scientific conclusions. It's a captivating idea, yet without verifiable evidence, it remains firmly in the realm of conjecture.

The **Rare Earth Hypothesis** focuses on the specific conditions required for the emergence of complex life on Earth. While the unique combination of factors is undeniable, arguing that this combination is uniquely rare requires a greater understanding of planetary formation and evolution across a statistically significant sample size. We currently lack the data to confidently extrapolate from our single data point—Earth—to the vast expanse of the universe. While Earth's unique characteristics might make it less common, it doesn't definitively prove that complex life is inherently rare. The discovery of even a single instance of complex life on another planet would dramatically weaken this hypothesis.

Our limited understanding of **life itself** poses a significant limitation in our search for extraterrestrial intelligence. Our search methods are inherently biased towards carbon-based life forms similar to those found on Earth. The possibility of silicon-based life or other exotic biochemistries opens up a universe of possibilities that our current methodologies may be entirely inadequate to detect. Furthermore, the timescale of civilization rise and fall, relative to the vastness of cosmic time, is another crucial variable. The short-lived nature of civilizations could easily explain the lack of detection, despite the potential abundance of past or future civilizations. The "overlapping" period of detectable signals might be incredibly short compared to the overall lifespan of the universe.

The limitations of our **current technology** are arguably the most tangible aspect to consider. Our search methods are limited, both in scope and sensitivity. We're essentially listening for signals in a specific range of frequencies, assuming a specific form of communication that might be entirely outmoded or irrelevant to a more advanced civilization. Our understanding of potential signals could be fundamentally flawed; we may be missing crucial information, misinterpreting data, or simply failing to recognize advanced forms of communication that we are not equipped to detect. The development of more sophisticated search strategies, broader frequency ranges, and more advanced data analysis techniques is therefore crucial to progress in the search for extraterrestrial life.

In conclusion, while the Fermi Paradox remains a significant enigma, the proposed resolutions are not mutually exclusive. Multiple factors may be at play, contributing to the apparent silence. The Great Filter could be operating in conjunction with the difficulties of interstellar travel, compounded by our limited technological capabilities and a narrow understanding of the diversity of life forms. The Rare Earth Hypothesis adds another layer of complexity, suggesting that even if life is common, complex life may be exceptionally rare. The zoo hypothesis provides an interesting philosophical counterpoint but lacks tangible evidence to support it. The challenge lies in designing and implementing more effective search strategies, expanding our understanding of the fundamentals of life and intelligence, and acknowledging the limitations of our current perspective. The pursuit of answers to the Fermi Paradox continues to drive research in astrobiology and pushes the boundaries of scientific inquiry, reminding us of the vastness of the universe and the immensity of the unknown. The ongoing search is a testament to our enduring curiosity and our relentless quest to understand our place in the cosmos. The silence may yet speak volumes, once we learn to truly listen.

The Fermi Paradox, in its elegant simplicity, forces us to confront the profound chasm between what we expect and what we observe. The sheer vastness of the universe, coupled with the estimated age of billions of years, suggests a high probability of extraterrestrial civilizations. Yet, despite decades of dedicated searches, we remain without definitive proof of their existence. This discrepancy, this uncomfortable silence, is the core of the paradox, and its significance extends far beyond the realm of scientific inquiry. It challenges our fundamental assumptions about the universe and our place within it, prompting a reevaluation of our anthropocentric worldview.

One of the most significant impacts of the Fermi Paradox is its effect on our understanding of the rarity of life. The prevalent assumption in the past has often been that Earth, with its flourishing biodiversity, represents a common occurrence in the cosmos. The apparent absence of detectable extraterrestrial civilizations, however, casts doubt on this assumption. It prompts us to consider the possibility that the emergence of life, particularly intelligent life, might be an exceptionally rare event, a cosmic anomaly rather than the norm. This re-evaluation necessitates a deeper exploration of the specific conditions that allowed life to flourish on Earth, prompting renewed research into astrobiology and the study of extremophiles – organisms that thrive in extreme environments – which may offer clues about the adaptability and resilience of life under challenging conditions. The conditions on Earth, including the presence of liquid water, a stable climate, and a protective magnetic field, are often cited as unique factors that may not be replicated across other planets. The paradox thus encourages a more nuanced and cautious approach to estimating the prevalence of life throughout the universe, forcing us to reconsider what constitutes a "habitable" zone and the diverse forms life might take.

Furthermore, the Fermi Paradox compels us to re-evaluate our assumptions about technological advancement. The assumption underlying many searches for extraterrestrial intelligence (SETI) is that other civilizations would develop technologies similar to our own, resulting in detectable signals. The paradox, however, suggests that this assumption might be flawed. Advanced civilizations may have developed technologies and communication methods far beyond our current comprehension, rendering our search strategies ineffective. They may have transitioned to forms of energy production and communication that are undetectable by our current methods, or they may have advanced beyond the need for broadcasting signals across vast interstellar distances. Consider, for example, the rapid pace of technological advancement on Earth. In just a few centuries, we've transitioned from rudimentary tools to sophisticated computing technologies. Extrapolating this trend to a civilization thousands or millions of years ahead of us suggests the possibility of technologies so radically different from our own that their existence would be entirely invisible to us. The paradox, therefore, challenges us to broaden our search strategies, exploring alternative methods of detection and reconsidering the very definition of "advanced technology."

The philosophical implications of the Fermi Paradox are perhaps even more profound. The apparent absence of contact compels us to confront our own assumptions about our place in the cosmos. For centuries, humankind has grappled with its place in the universe, oscillating between the belief that we are unique and the possibility that we are not alone. The Fermi Paradox intensifies this contemplation, compelling us to question the inherent human-centric bias in our understanding of the universe. Are we truly significant, or are we merely a fleeting, insignificant blip in the cosmic timeline? The silence of the universe, in the context of the paradox, can be interpreted as both a humbling and terrifying reality.

The paradox also prompts us to consider the potential dangers of advanced technologies. If civilizations do exist, but are absent from the cosmic conversation, it raises the possibility of self-destruction or some form of catastrophic technological failure. The development of advanced technologies, particularly those involving immense energy resources, might harbor inherent risks that could lead to the demise of a civilization before it can achieve interstellar communication. This sobering perspective underscores the importance of responsible technological development on Earth and highlights the urgency of addressing potential global threats, such as nuclear war, climate change, and artificial intelligence. The Fermi Paradox serves as a cautionary tale, reminding us of the potential fragility of civilization and the importance of navigating the complexities of technological progress carefully.

Moreover, the implications extend to our understanding of time. The vast scales of cosmic time are difficult for humans to fully comprehend. A civilization might rise and fall within a relatively short period, leaving behind no trace for us to detect. Consider that the lifespan of a star is measured in billions of years, and the distances between stars are measured in light-years. A civilization's window of detectability might be incredibly brief in the grand scheme of the universe. They might exist, flourish, and vanish without ever leaving a detectable signal, or their signal could simply be too faint or too distant to reach us. This consideration emphasizes the vastness of the universe and challenges us to acknowledge the limitations of our temporal perspective.

The Fermi Paradox also highlights the limitations of our current scientific understanding. Our current methods for detecting extraterrestrial life are based on our limited understanding of biology, physics, and communication. We are searching for life that is similar to life as we know it, which may be a fundamental flaw. Extraterrestrial life might exist in forms that are utterly alien to our current comprehension, rendering our search methods ineffective. It's also possible that our understanding of physics and communication is incomplete, limiting our ability to detect signals from civilizations using technologies far beyond our current grasp. We are effectively listening for radio waves while perhaps ignoring other, more subtle or sophisticated communication methods entirely.

In conclusion, the Fermi Paradox is not simply a scientific puzzle; it is a profound philosophical challenge that prompts us to reconsider our assumptions about the universe, our place within it, and the very nature of life and intelligence. The silence is not simply an absence of signals; it's a rich source of information prompting us to improve our methods, expand our understanding, and confront the possibilities of both extraordinary discovery and unsettling reality. It continues to drive research in astrobiology, cosmology, and other related fields, pushing the boundaries of our current understanding and reminding us of the vastness and mystery of the cosmos. The search for extraterrestrial intelligence is, ultimately, a search for a deeper understanding of ourselves and our place in the grand tapestry of existence. The paradox remains, yet the pursuit of its resolution fuels our scientific endeavor and fosters profound philosophical reflection. The silence, in its enigmatic profundity, compels us to continue searching, to learn to listen more effectively, and to prepare for the unexpected.

The Fermi Paradox, with its unsettling juxtaposition of expectation and observation, continues to drive scientific inquiry and shape our understanding of the cosmos. Addressing this paradox requires a multi-pronged approach, encompassing advancements in several key research areas. Ongoing research is focused on refining our search strategies, expanding the scope of our investigations, and developing more sophisticated theoretical frameworks to interpret the available data.

One of the most significant ongoing research efforts involves the improvement of our methods for detecting extraterrestrial intelligence (SETI). Traditional SETI programs have largely focused on radio signals, a technology relatively short-lived in the context of a civilization's potential lifespan. This approach is based on the assumption that other civilizations would use similar technologies for communication. The Fermi Paradox itself challenges this anthropocentric assumption, suggesting that advanced civilizations might employ methods of communication beyond our current comprehension, possibly utilizing forms of energy or information transfer that we cannot yet detect. As such, current research is expanding to incorporate more diverse detection methods. This includes the search for technosignatures, which are the byproducts of advanced technologies detectable from afar, such as unusual energy emissions, atmospheric anomalies suggestive of industrial activity, or even megastructures orbiting distant stars. The development of more sensitive and sophisticated instruments, combined with advanced data analysis techniques, is crucial to this endeavor. Projects are underway to analyze astronomical data for evidence of Dyson spheres, hypothetical megastructures built around stars to harness their energy, or other indications of advanced civilizations' engineering capabilities. These searches rely heavily on advanced computational techniques and machine learning algorithms to sift through vast quantities of astronomical data, identifying potentially anomalous signals or patterns that might be missed by human observers.

Another vital aspect of ongoing research focuses on expanding our understanding of extremophiles and the limits of life. The discovery and study of organisms thriving in extreme environments on Earth – from deep-sea hydrothermal vents to highly acidic or radioactive environments – provides crucial insights into the potential for life to exist in seemingly inhospitable conditions across the universe. By understanding the adaptability and resilience of life on Earth, we can refine our models of habitability and broaden our search for extraterrestrial life beyond the traditional "Goldilocks zone" around stars. Understanding the range of conditions under which life can exist is vital for identifying potential biosignatures, or indicators of past or present life, on other planets and moons. This research incorporates advances in genomics, biochemistry, and planetary science, allowing for a more comprehensive understanding of the potential diversity of life beyond Earth and the conditions that may nurture it.

Theoretical research plays a critical role in addressing the Fermi Paradox. Scientists are exploring various hypotheses that attempt to resolve the apparent contradiction between the high probability of extraterrestrial life and the lack of observed contact. These hypotheses range from the "Great Filter" concept, which posits that there is a significant obstacle preventing the emergence or survival of intelligent life, to the "Zoo Hypothesis," which suggests that advanced civilizations are deliberately avoiding contact with less-developed civilizations like ours. Other hypotheses explore the possibility of civilizations existing in different dimensions or realities, making contact impossible by our current means of understanding. These theories require sophisticated modeling and simulations to explore their plausibility and potential implications, involving collaborative efforts between astrophysicists, mathematicians, and philosophers. Testing these hypotheses requires the development of innovative observational techniques and the incorporation of multidisciplinary perspectives.

Furthermore, our understanding of the timescale involved in cosmic events is crucial for interpreting the Fermi Paradox. The vast age of the universe, coupled with the relatively short lifespan of technologically advanced civilizations, means that the probability of overlapping existence – the period when two civilizations are both technologically advanced and capable of interstellar communication – might be remarkably small. It's possible that the sheer vastness of time and space makes the probability of detecting another civilization remarkably low, even if many exist. Research on this aspect involves advanced cosmological models and statistical analysis, attempting to quantify the probability of encountering another civilization based on various assumptions about the rate of star formation, the prevalence of life, and the average lifespan of technological civilizations.

Future directions in research involve ambitious initiatives like the development of advanced telescopes and space-based observatories. These will be capable of detecting fainter signals and a wider range of technosignatures, allowing for a more thorough search for evidence of extraterrestrial intelligence. The development of next-generation telescopes, such as the Extremely Large

Telescope (ELT) and the James Webb Space Telescope (JWST), will offer unprecedented capabilities in observing distant planets and stars, potentially uncovering evidence of atmospheric biomarkers or signs of technological activity. These advancements are essential in expanding the scope and sensitivity of our search efforts.

The integration of artificial intelligence and machine learning is also vital for future progress. These technologies will allow scientists to analyze massive datasets from astronomical observations, identifying subtle patterns or anomalies that might be missed by human analysis. Machine learning algorithms can be trained to recognize potential technosignatures, such as unusual light curves or radio emissions, helping to sift through vast quantities of data efficiently and potentially uncover evidence that would otherwise go unnoticed. This integration requires the collaboration of astronomers, computer scientists, and data scientists.

Finally, a crucial aspect of future research involves fostering interdisciplinary collaboration. The Fermi Paradox is not solely a scientific problem; it touches upon philosophical, ethical, and sociological aspects of human existence. Addressing this paradox requires a synthesis of knowledge from various fields, including astrobiology, cosmology, anthropology, sociology, and philosophy. Collaborative projects that involve researchers from diverse backgrounds are essential for developing innovative approaches and interpretations of the data. This interdisciplinary approach will be crucial for building a more comprehensive understanding of the implications of the Fermi Paradox and its relevance to the human experience.

In conclusion, the Fermi Paradox is not an insurmountable hurdle but rather a catalyst for scientific advancement and a profound invitation to expand our understanding of the universe and our place within it. Ongoing and future research efforts, fueled by technological innovation and interdisciplinary collaboration, hold the promise of uncovering answers to this enduring enigma. The quest to resolve the Fermi Paradox is not merely a search for extraterrestrial life but a journey of self-discovery, pushing the boundaries of human knowledge and challenging our deeply held assumptions about the universe. The pursuit of this answer compels us to continuously refine our methods, broaden our perspectives, and prepare for the astonishing and potentially transformative discoveries that lie ahead. The silence, however profound, fuels our ongoing exploration and the unwavering pursuit of answers.

| 8 |

Astrobiology and the Search for Life

Astrobiology, in its simplest definition, is the study of life's origins, evolution, distribution, and future in the universe. It's a field that transcends the boundaries of traditional scientific disciplines, embracing a holistic approach that integrates knowledge from astronomy, biology, chemistry, geology, and even planetary science. Instead of focusing on life solely as we know it on Earth, astrobiology explores the vast potential for life to exist in diverse forms and under a wider range of conditions than previously imagined. This interdisciplinary nature is at the very heart of the field, recognizing that the search for life beyond Earth requires a multifaceted approach that combines expertise from numerous areas of scientific inquiry.

Consider, for example, the search for habitable planets. Astronomers use sophisticated telescopes and observational techniques to identify exoplanets—planets orbiting stars other than our Sun—and characterize their atmospheres. The analysis of an exoplanet's atmosphere, a process that leverages the expertise of chemists and physicists, can reveal the presence or absence of key biosignatures—gases or other substances that could indicate the presence of life. Geologists, meanwhile, contribute their understanding of planetary formation and evolution, providing insights into the potential for a planet to sustain liquid water, a crucial component for life as we know it. Furthermore, understanding the geological processes that shape a planet's surface, its internal structure, and its interaction with its atmosphere is crucial to assessing its potential habitability. This interdisciplinary effort is essential to properly evaluate and interpret the vast amounts of data generated by astronomical observations and sophisticated laboratory analysis.

The biological aspects of astrobiology extend beyond simply searching for life; it also investigates the limits of life itself. The discovery of extremophiles—organisms that thrive in extreme environments on Earth, such as deep-sea hydrothermal vents, acidic hot springs, or highly radioactive areas—has profoundly impacted our understanding of the potential for life to exist in seemingly inhospitable places in the universe. These extremophiles, studied by biologists and biochemists, highlight the remarkable adaptability and resilience of life. They challenge our anthropocentric view of what constitutes a habitable environment, broadening our search for extraterrestrial life beyond the traditional "Goldilocks zone"—the region around a star where conditions are considered just right for liquid water to exist on a planet's surface. The study of extremophiles allows us to refine our models of habitability, considering a much wider range of environmental conditions that could support life. This includes considering the possibility of life in subsurface oceans on icy moons, environments shielded from harsh radiation, and places where unique chemical processes could sustain life independent of sunlight.

The goals of astrobiology are ambitious and far-reaching. Ultimately, it strives to answer fundamental questions about the origin and prevalence of life in the universe. These questions encompass several key areas of investigation:

The Origin of Life: Astrobiology seeks to understand how life arose on Earth and whether similar processes might have occurred elsewhere. This involves investigating prebiotic chemistry—the chemical reactions that led to the formation of the building blocks of life—and exploring the conditions necessary for the transition from non-living matter to living organisms. Laboratory experiments that simulate early Earth conditions are employed to recreate these processes and test hypotheses regarding the origin of life. The search for prebiotic molecules in meteorites and on other celestial bodies provides further clues to the potential distribution of life's building blocks throughout the universe.

The Habitability of Other Worlds: Astrobiology assesses the potential for life to exist on other planets and moons within our solar system and beyond. This involves characterizing the environments of these celestial bodies, considering factors such as the presence of liquid water, suitable temperatures, and the availability of essential elements for life. The development of advanced robotic missions and future human space exploration will play a crucial role in collecting samples and performing in-situ experiments to directly investigate the habitability of other worlds. The search for biosignatures, or indicators of past or present life, in the atmospheres of exoplanets is also a key aspect of determining their potential habitability.

The Diversity of Life: Astrobiology investigates the potential diversity of life beyond Earth. Given the remarkable diversity of life on Earth, it is plausible that life elsewhere might exhibit vastly different forms and biochemical processes. The study of extremophiles on Earth helps to expand our understanding of the potential for life to exist under diverse environmental conditions. The search for extraterrestrial life will likely involve detecting life that differs significantly from life on Earth, and thus requires development of sensitive instruments and methodologies.

The Future of Life: Astrobiology explores the long-term evolution and future of life, both on Earth and elsewhere in the universe. This includes investigating the potential for life to adapt to changing environmental conditions, the long-term sustainability of life, and the potential impact of human activities on the evolution of life on Earth and beyond. The research is critical to gaining a perspective on our own planet's future and developing strategies for planetary protection and preservation of ecosystems.

The interdisciplinary nature of astrobiology is exemplified by its close collaboration with other fields. For instance, the search for extraterrestrial life often leverages advancements in astronomy, with improved telescopes and observational techniques leading to the discovery of new exoplanets and the characterization of their atmospheres. Chemistry plays a crucial role in understanding the building blocks of life and the chemical processes that might support life under diverse conditions. Geology provides insights into the formation and evolution of planets and the potential for subsurface oceans or other habitats. Biologists are vital to comprehending the adaptability of life and defining the limits of habitability. Finally, the integration of data analysis and modeling techniques, driven by the advances in computer science and mathematics, allows for the processing of massive amounts of astronomical data, the identification of potential biosignatures, and the creation of sophisticated simulations of planetary environments and the origin of life.

The study of astrobiology is not just about searching for alien life, though that is a central component. It's also about gaining a deeper understanding of our own planet, its history, and its future. By studying the limits of life and the conditions under which it can thrive, we can gain a more profound appreciation of the interconnectedness of all living things and the fragility of ecosystems. It challenges us to consider our place within the universe, not as a unique and isolated species but as part of a larger cosmic story, one that may yet reveal incredible wonders and profound implications for the future of humanity. The ongoing quest to understand life's origins, evolution, and distribution within the vast cosmos is a testament to human curiosity and our relentless drive to explore the unknown. The field of astrobiology, in its multidisciplinary approach, is at the forefront of this exploration, constantly pushing the boundaries of scientific knowledge and shaping our understanding of our place in the universe.

The search for extraterrestrial life hinges on a robust theoretical framework, a set of guiding principles and concepts that shape our investigations and interpret our findings. This framework isn't static; it evolves as our understanding of life, both on Earth and potentially elsewhere, deepens. Understanding these core concepts is crucial to appreciating the complexities and nuances of astrobiological research.

One fundamental concept is **habitability**, the capacity of an environment to support life. This is not simply the presence of liquid water, though that's a critical factor for life as we know it. Habitability is a multifaceted concept encompassing a range of parameters, including:

Availability of liquid water: While liquid water is essential for many known biological processes, the search for life isn't confined to environments with water at Earth-like temperatures and pressures. The discovery of extremophiles on Earth, thriving in sub-zero temperatures, highly acidic conditions, or intense radiation, has dramatically expanded our understanding of what constitutes a habitable environment. Subsurface oceans on icy moons like Europa and Enceladus, for instance, are considered potential habitats despite the frigid surface temperatures. Liquid water, in various forms and under various conditions, is a key focus.

Presence of essential elements: Life requires certain chemical building blocks, primarily carbon, hydrogen, oxygen, nitrogen, phosphorus, and sulfur (CHNOPS). The abundance and accessibility of these elements are crucial for the emergence and sustenance of life. While carbon-based life is the only form we currently know, the possibility of silicon-based or other forms of life cannot be discounted. Astrobiological research therefore involves not just searching for CHNOPS, but investigating the availability and potential reactivity of a broader spectrum of elements under diverse conditions.

Energy sources: Life requires an energy source to drive metabolic processes. On Earth, this is primarily sunlight, but other sources, like hydrothermal vents, geothermal energy, or chemical reactions, also support life. When evaluating potential habitats, researchers must consider the availability of diverse energy sources, beyond the familiar solar radiation, to expand the search for potentially unique forms of life.

Environmental stability: A stable environment, at least over a significant period, is necessary for life to evolve and flourish. Dramatic changes in temperature, atmospheric pressure, radiation levels, or other environmental factors could be detrimental. Understanding the long-term stability of a planetary environment is therefore crucial to assessing its habitability. This involves studying planetary geology, climate history, and the potential for catastrophic events that could render a planet uninhabitable.

Closely tied to habitability is the concept of **biosignatures**, indicators of past or present life. These can take various forms:

Biological molecules: The detection of organic molecules, such as amino acids, nucleotides, or lipids, could be strong evidence of past or present life. However, the abiotic formation of these molecules must be carefully considered. Sophisticated analytical techniques are crucial to distinguish between biologically produced and naturally occurring organic matter.

Gases in planetary atmospheres: Certain gases in a planet's atmosphere could be indicative of biological activity. For example, the presence of oxygen, methane, or other gases at unusual concentrations could suggest biological processes. However, the interpretation of atmospheric gases must consider a range of factors, including geological activity and atmospheric chemistry.

Geological formations: Unique geological formations could indicate the presence of past microbial life. Stromatolites, layered sedimentary formations produced by microbial mats, are a prime example of biosignatures found on Earth. Identifying similar structures on other planets could indicate past life.

Technological signatures (technosignatures): This refers to evidence of advanced technology created by an extraterrestrial civilization, such as radio waves, laser signals, or megastructures detectable from afar. The search for technosignatures is a relatively new but increasingly important area of astrobiological research.

The concept of **evolution** is central to astrobiology. Understanding the evolutionary processes that shaped life on Earth provides a framework for predicting the potential diversity of life elsewhere. Key principles of evolution include:

Natural selection: The process by which organisms with traits better suited to their environment are more likely to survive and reproduce. This leads to the adaptation of organisms to their surroundings over time.

Genetic variation: Variation within a population provides the raw material for natural selection. This variation can arise from mutations, gene flow, or sexual reproduction.

Common ancestry: The theory that all life on Earth shares a common ancestor. This implies a shared evolutionary history and potential commonality in fundamental biological processes.

Applying these principles to extraterrestrial life suggests that, given diverse environmental conditions, life elsewhere might exhibit an equally diverse range of adaptations and evolutionary pathways. The study of extremophiles on Earth demonstrates the surprising resilience and adaptability of life, expanding the range of possible habitats and life forms that might exist in the universe. This underlines the necessity of expanding our search beyond environments mirroring Earth's conditions.

Furthermore, the concept of **panspermia**, the hypothesis that life exists throughout the Universe and is distributed by meteoroids, asteroids, and comets, challenges our understanding of the origin and distribution of life. If life can be transferred between planets, even across vast interstellar distances, then the possibility of life arising independently on multiple planets is increased significantly, potentially even leading to a "cosmic web of life" connecting biological systems across the galaxy.

Finally, the ethical considerations of discovering extraterrestrial life are significant. The potential impact on society, religion, philosophy, and science would be profound. Astrobiological research must incorporate these ethical considerations into its methodology and future plans. The potential for contamination of extraterrestrial environments by terrestrial life or vice versa demands stringent protocols for planetary protection. These considerations underscore the responsibility of scientific communities to approach such discoveries thoughtfully and responsibly.

In conclusion, the key concepts of habitability, biosignatures, and evolution provide a framework for understanding the search for life beyond Earth. Understanding these principles, as well as the associated ethical considerations, is crucial for interpreting data, designing experiments, and ultimately, understanding our place within the vast cosmic landscape. The pursuit of astrobiology is not just a scientific endeavor; it is a voyage of discovery into the deepest questions of existence, driving innovation and reshaping our understanding of life itself.

The theoretical framework laid out in the previous section provides the foundation for numerous ongoing astrobiological research projects. These projects employ a variety of approaches, from robotic exploration of distant celestial bodies to sophisticated laboratory experiments designed to simulate extraterrestrial environments. The overarching goal remains consistent: to identify past or present life beyond Earth, or at the very least, to better understand the conditions that could support life elsewhere.

One of the most ambitious and highly publicized efforts is the ongoing exploration of Mars. Several robotic missions, including the Perseverance rover and the Ingenuity helicopter, are currently actively searching for biosignatures on the Martian surface. Perseverance, equipped with advanced instruments capable of analyzing rock and soil samples, is collecting samples for potential future return to Earth. These samples will be subjected to even more rigorous analysis than is possible in situ, offering the potential to uncover microscopic fossils or other indicators of past life. The Ingenuity helicopter has expanded the operational range of the rover, enabling exploration of previously inaccessible terrains. Data from these missions, coupled with observations from orbiters like the Mars Reconnaissance Orbiter (MRO) and the Trace Gas Orbiter (TGO), which scrutinize the Martian atmosphere for trace gases indicative of biological activity, are providing an increasingly comprehensive picture of Mars' past and present habitability. The search is not just focused on the surface; subsurface exploration is also considered a high priority, given the possibility of liquid water existing beneath the Martian surface. Future missions may involve drilling deep beneath the surface to access these potentially habitable subsurface environments.

Beyond Mars, other celestial bodies within our own solar system are attracting considerable attention. Jupiter's moon Europa is a prime example. Strong evidence suggests a vast subsurface ocean, potentially containing more water than all of Earth's oceans combined, hidden beneath a thick layer of ice. The planned Europa Clipper mission by NASA aims to conduct a detailed reconnaissance of Europa, using a suite of instruments to study the composition of the ice shell, the subsurface ocean, and the moon's tenuous atmosphere. The Clipper will map Europa's surface, analyze its plumes (if any are detected), and measure the thickness of the ice shell, all in an effort to assess the habitability of the subsurface ocean and search for signs of life. Similar missions are being planned for Enceladus, another icy moon orbiting Saturn, which also shows evidence of subsurface water and geysers erupting from its south polar region. The Cassini mission provided some tantalizing data on Enceladus's plumes, and future missions will aim to study these plumes more comprehensively to analyze their chemical composition for potential biosignatures. The search for life on these icy moons requires innovative technologies that can penetrate the ice and analyze the ocean's contents remotely, a challenging but critically important technological hurdle.

Our understanding of extremophiles on Earth continues to inform and expand the search for extraterrestrial life. The discovery of life thriving in environments previously thought to be uninhabitable—from the deepest ocean trenches to the driest deserts—broadens the range of possibilities for where life might exist elsewhere. Laboratory experiments simulating extreme environments, such as those found on Mars, Europa, or Enceladus, are being conducted to test the limits of life's resilience and to better understand the conditions that might support life beyond our current understanding. These experiments involve cultivating extremophiles under simulated extraterrestrial conditions, studying their metabolic pathways, and testing their survival limits. The goal is to understand which biological processes can withstand the intense radiation, extreme temperatures, and other challenging conditions found on other planets and moons. This research helps to refine our search strategies, by identifying specific biosignatures that might be more readily detectable in these extreme environments.

Furthermore, the search extends beyond our solar system. The Kepler space telescope and its successor, the Transiting Exoplanet Survey Satellite (TESS), have discovered thousands of exoplanets, planets orbiting stars other than our Sun. Characterizing the atmospheres of these exoplanets is a major focus of current research. Using advanced spectroscopic techniques, astronomers are analyzing the light passing through these atmospheres to search for biosignatures such as oxygen, methane, or other gases that could indicate the presence of life. This is a challenging undertaking, requiring highly sensitive instruments and sophisticated data analysis techniques. The James Webb Space Telescope, with its unprecedented sensitivity, is poised to play a critical role in this endeavor, offering unparalleled opportunities to analyze the atmospheres of exoplanets in greater detail than ever before. Identifying potentially habitable exoplanets is just the beginning; future missions will need to be designed to further study these planets, ideally with advanced telescopes and potentially with interstellar probes, to more thoroughly investigate the possibilities of life beyond our solar system.

The search for technosignatures, evidence of advanced extraterrestrial civilizations, is also an active area of research. Projects like the Breakthrough Listen initiative are using powerful radio telescopes to scan the skies for artificial signals that might originate from extraterrestrial civilizations. This involves analyzing vast amounts of radio data for unusual patterns or signals that could not be easily attributed to natural phenomena. The search for technosignatures also includes looking for megastructures, large-scale constructions created by extraterrestrial civilizations that might be detectable from afar. These could include Dyson spheres, hypothetical megastructures built around stars to harness their energy. The detection of technosignatures would represent a monumental discovery, potentially altering our understanding of the universe and our place within it.

The ethical considerations associated with the search for extraterrestrial life are also increasingly being incorporated into the planning and execution of astrobiological research. The potential discovery of extraterrestrial life raises a multitude of complex ethical questions regarding planetary protection (preventing contamination of other celestial bodies with terrestrial life), the responsible disclosure of findings, and the potential impacts on society, religion, and philosophy. International collaborations and guidelines are essential to ensuring a responsible and ethical approach to this potentially transformative scientific endeavor. The development of robust protocols for planetary protection is crucial to avoid contaminating potentially habitable environments with terrestrial microorganisms, potentially jeopardizing the search for extraterrestrial life and introducing biases into our analysis.

In summary, the search for extraterrestrial life is a multifaceted and dynamic endeavor involving a global community of researchers. From robotic missions to distant celestial bodies to advanced laboratory experiments and sophisticated data analysis techniques, the pursuit of astrobiology is driving scientific innovation and pushing the boundaries of our understanding of life itself. The ongoing and planned missions described above represent only a fraction of the diverse research projects underway. Each project contributes a vital piece to the puzzle, gradually increasing our knowledge and advancing our ability to answer the profound question: Are we alone? The answer, though still elusive, draws closer with every passing year, propelled by relentless scientific curiosity and innovative technologies. The ongoing discoveries and refinement of methodologies will likely lead to even more intriguing and ambitious projects in the coming decades. The journey into the unknown, fueled by decades of research and the excitement of discovery, continues, promising to reshape our understanding of the universe and humanity's place within it.

The quest to understand if life exists beyond Earth hinges critically on technological advancements. Without the relentless innovation in instrumentation and data analysis, astrobiology would remain largely a field of theoretical speculation. The progress made in recent decades is nothing short of astonishing, transforming the search from a realm of science fiction to a rigorous and data-driven scientific endeavor.

One of the most transformative technological advancements has been the development of increasingly powerful and sophisticated telescopes. The Hubble Space Telescope, launched in 1990, revolutionized our understanding of the universe, providing unprecedented views of distant galaxies and nebulae. Its observations have been instrumental in characterizing exoplanet atmospheres, providing clues about their potential habitability. However, Hubble's capabilities are now being surpassed by the James Webb Space Telescope (JWST), launched in 2021. JWST's infrared capabilities allow it to peer through dust clouds, providing unparalleled views of star formation and enabling the detailed spectroscopic analysis of exoplanet atmospheres. This allows for the detection of biosignatures – specific chemical compounds or patterns in an exoplanet's atmosphere that could indicate the presence of life. The sensitivity of JWST is several orders of magnitude greater than its predecessors, enabling the detection of trace gases that might have been missed by previous instruments. This means we can now detect smaller, fainter planets and analyze their atmospheres with far greater precision, pushing the boundaries of what we can learn about planets orbiting other stars.

Beyond optical telescopes, radio telescopes play a crucial role in the search for technosignatures – evidence of advanced extraterrestrial civilizations. The Allen Telescope Array (ATA), a collection of hundreds of radio antennas, is dedicated to searching for radio signals that might originate from extraterrestrial civilizations. The ATA's large collecting area and wide field of view allow it to survey vast swathes of the sky with unprecedented speed and efficiency. The Breakthrough Listen initiative, utilizing some of the world's most powerful radio telescopes, builds on this effort, employing sophisticated signal processing techniques to sift through the vast amounts of data collected. The development of advanced algorithms and computational techniques is vital here, allowing researchers to filter out noise and identify potentially interesting signals that would otherwise be lost in the background noise of the universe. Further advancements in the design and sensitivity of radio telescopes are crucial to expanding the search for these technosignatures, particularly as many researchers suspect that radio signals may be just one aspect of an advanced civilization's technological output.

The exploration of our solar system relies heavily on advancements in spacecraft technology. Robotic missions are essential for the in-situ exploration of planets and moons, as human missions to these distant destinations are still far in the future, technologically and financially. The Mars rovers, like Curiosity and Perseverance, are examples of sophisticated robotic explorers capable of traversing the Martian surface, collecting samples, and performing scientific analysis. These rovers are equipped with a wide array of instruments, including spectrometers, cameras, and drills, enabling them to analyze the composition of rocks and soil and search for biosignatures. The Ingenuity helicopter, a technological marvel in itself, has demonstrated the feasibility of aerial exploration of Mars, enabling access to areas previously unreachable by rovers. These advancements in robotics are not confined to Mars; similar technology is being developed for missions to other celestial bodies within our solar system, including Europa and Enceladus, which possess subsurface oceans that may harbor life.

The technological challenges of exploring these icy moons are immense. Future missions will require innovative technologies capable of penetrating the thick ice shells and analyzing the composition of these subsurface oceans. This might involve the use of advanced radar systems, powerful drills, or even submersible vehicles capable of operating in extreme environments. The development of autonomous systems, capable of functioning independently in harsh conditions, will also be critical for these missions. These autonomous systems will be critical to navigate unpredictable terrain, conduct scientific experiments without human intervention, and make real-time decisions based on the data they collect. The advancements in autonomous navigation are being tested in various domains, such as underwater exploration and deep space navigation, and the future promises further advancements in this key field.

Beyond spacecraft, significant advancements have been made in analytical instruments, vital for analyzing samples collected from space missions. Mass spectrometers, for example, can identify the isotopic composition of molecules, providing insights into the processes that formed them. Sophisticated techniques, like gas chromatography-mass spectrometry (GC-MS) and liquid chromatography-mass spectrometry (LC-MS), allow researchers to analyze complex mixtures of organic compounds, searching for unique biosignatures. The development of miniature, lightweight instruments, capable of operating in extreme conditions, is crucial for space exploration. These instruments must be robust, reliable, and capable of producing high-quality data despite the limitations imposed by operating in harsh environments with limited power and communication capabilities. Microscopy techniques are also undergoing constant improvements, with increased resolution and sensitivity, providing ever more powerful tools for detecting microscopic organisms or fossils.

Furthermore, the analysis of data obtained from these missions requires increasingly sophisticated computational techniques. The sheer volume of data generated by modern spacecraft and telescopes necessitates the development of advanced algorithms and high-performance computing capabilities. Machine learning techniques are being used to automate data analysis, identify patterns, and potentially discover anomalies that might indicate the presence of life. The ability to efficiently and effectively handle this vast and complex dataset is rapidly becoming a limiting factor, underscoring the importance of developing faster, more powerful algorithms and hardware to keep pace with the data flow. The convergence of advanced computing, artificial intelligence, and high-throughput data analysis is critical to extracting meaning from the wealth of information being generated in the pursuit of astrobiology.

Finally, advancements in materials science are playing a crucial role in enabling the development of instruments capable of withstanding the harsh conditions encountered during space exploration. The development of radiation-hardened electronics, high-temperature materials, and robust sensor technologies is essential for the success of future missions. These materials must be lightweight, durable, and capable of withstanding extreme temperature variations, radiation, and other environmental stresses. The development of new materials is an ongoing effort, constantly pushing the boundaries of what is possible. Innovative materials, and their ability to withstand the rigors of space travel, are continuously improving, enabling the creation of more sophisticated and reliable instruments that will enable more ambitious and far-reaching explorations.

In conclusion, the technological advancements discussed above are not merely supporting elements but are the very driving force behind the progress in astrobiology. From advanced telescopes capable of peering into the atmospheres of distant exoplanets to robotic spacecraft exploring the surfaces of other worlds and sophisticated analytical instruments capable of detecting faint biosignatures, technology is continually pushing the boundaries of our ability to answer the fundamental question of whether life exists beyond Earth. The future of astrobiology is inextricably linked to the continued development of these technologies and the innovative ways in which they are applied. The pace of technological advancement suggests that we are on the cusp of a new era of discovery, potentially revealing answers to questions that have captivated humanity for millennia.

The rapid pace of technological advancement promises a future where the search for extraterrestrial life transitions from a largely theoretical pursuit to one dominated by empirical evidence. Several key areas hold the potential for transformative breakthroughs in the coming decades. One of the most promising is the development of next-generation telescopes. While the James Webb Space Telescope represents a monumental leap forward, its capabilities will inevitably be surpassed. Future space-based observatories could incorporate technologies like advanced adaptive optics to compensate for atmospheric distortion, allowing for even sharper images and more detailed spectroscopic analysis of exoplanet atmospheres. This could lead to the detection of increasingly subtle biosignatures, possibly including indicators of complex life or even technological activity. The development of extremely large telescopes on the ground, utilizing interferometry techniques to combine the light from multiple telescopes, also holds great potential for enhancing our observational capabilities. These advances will allow us to probe the atmospheres of smaller, potentially Earth-like planets with unprecedented precision, significantly improving our chances of finding definitive evidence of life.

Furthermore, the search for technosignatures will likely benefit from advancements in radio astronomy, as well as the exploration of other potential signals. Beyond simply searching for radio waves, researchers are exploring the potential of detecting other forms of technological leakage, such as laser emissions, or even subtle changes in the light curves of stars that might indicate the presence of megastructures. The development of more sophisticated algorithms and artificial intelligence (AI) will be crucial for analyzing the immense amount of data generated by these searches. AI can help to filter out noise, identify patterns that might be missed by human observers, and prioritize potentially interesting signals for further investigation. The integration of AI into the data analysis pipeline will be critical for managing the ever-increasing volume of data from both ground-based and space-based observatories. The ongoing development of quantum computing may also offer unprecedented opportunities for signal processing and data analysis, potentially allowing us to detect extremely faint or complex signals currently beyond our reach.

The exploration of our solar system will continue to play a crucial role in the search for life beyond Earth. Future missions to Mars will likely focus on searching for fossilized evidence of past life, as well as investigating the potential for extant microbial life in subsurface environments. This will require the development of more sophisticated robotic explorers, capable of drilling deeper into the Martian subsurface and collecting samples with minimal contamination. Furthermore, advanced sample-return missions are essential for bringing Martian samples back to Earth for detailed analysis in sophisticated terrestrial laboratories. This process involves overcoming numerous challenges, including the development of robust containment systems to prevent contamination of the Earth's environment by any potential Martian lifeforms. The technological hurdles are significant, but the potential rewards – the ability to study Martian samples with the most powerful instruments available on Earth – are immense.

Beyond Mars, Europa and Enceladus, icy moons of Jupiter and Saturn respectively, represent compelling targets in the search for life. These moons possess subsurface oceans that may harbor life, shielded from the harsh radiation environments on their surfaces. Future missions to these icy worlds will require the development of innovative technologies capable of penetrating the thick ice shells and sampling the subsurface oceans. This could involve the use of advanced radar systems to map the subsurface oceans, specialized drills to penetrate the ice, and autonomous submersible vehicles capable of operating in the extreme conditions of these alien oceans. Developing reliable power sources, robust communication systems, and advanced navigation technologies will be essential for the success of these missions. The scientific instruments used will also need to be highly specialized, capable of detecting the unique biosignatures that might be found in these extreme environments.

The study of extremophiles on Earth continues to inform our search for life beyond Earth. Extremophiles are organisms that thrive in environments considered hostile to most life forms, such as extreme temperatures, pressures, or pH levels. Studying these organisms helps us to expand our understanding of the limits of life and identify potential biosignatures that could be found on other planets. The development of new genomic techniques, such as metagenomics, allows us to analyze the genetic material of entire microbial communities, providing insights into the diversity and metabolic capabilities of extremophiles. This information, in turn, informs our search for similar organisms in other environments, both on Earth and on other planets. Further research into extremophiles will be crucial to refining our search strategies and improving our ability to identify potential biosignatures in extraterrestrial samples.

Advancements in artificial intelligence and machine learning will revolutionize various aspects of astrobiological research. AI can assist in the design of experiments, the analysis of large datasets, and the development of new hypotheses. AI algorithms can sift through vast amounts of spectroscopic data to identify potential biosignatures that might be missed by human observers. They can also be used to control and operate robotic explorers autonomously, enabling exploration of hazardous environments where human intervention is impractical. The increasing sophistication of AI allows us to develop more comprehensive models of planetary environments and to better understand the factors that influence the origin and evolution of life. This also extends to the modeling of exoplanet atmospheres, allowing us to better assess their habitability and predict the types of biosignatures we might expect to find.

The development of new materials science is critical for future space exploration. Future missions will require instruments and spacecraft capable of withstanding the extreme conditions encountered during deep space travel. This necessitates the development of radiation-hardened electronics, lightweight and high-strength materials, and robust sensor technologies. These materials will need to be able to withstand extreme temperature fluctuations, high levels of radiation, and the vacuum of space. Further research into advanced materials will be essential for developing more resilient and sophisticated instruments for future space missions. Moreover, understanding the limits of material resilience in extreme environments will help design missions tailored to those specific challenges and, importantly, help analyze data from missions undertaken in these extreme environments. Furthermore, this development would help interpret data more accurately.

Finally, international collaboration will be crucial for accelerating progress in astrobiology. The sheer scale of the challenges involved, from designing and building sophisticated spacecraft to analyzing vast datasets, requires a concerted effort from scientists and engineers across the globe. Sharing resources, data, and expertise will be essential for maximizing efficiency and accelerating the pace of discovery. International collaborations not only leverage diverse skillsets and resources but also foster a sense of shared global responsibility in answering one of humanity's most fundamental questions. This shared responsibility will ultimately lead to more focused, efficient, and far-reaching exploration and research. Open-access data sharing and collaboration platforms will be critical for fostering this kind of international collaboration.

In summary, the future of astrobiology is bright. The combination of technological advancements, innovative research strategies, and international collaboration promises a new era of discovery. While the challenges are immense, the potential rewards – understanding our place in the universe and potentially discovering life beyond Earth – are immeasurable. The journey to answer the fundamental question of whether we are alone in the universe is a long and challenging one, but the progress made so far, and the trajectory of future research, suggest that we are on the cusp of potentially answering this age-old question. The coming decades promise an exciting and transformative period for astrobiology, with the potential for groundbreaking discoveries that will reshape our understanding of life and the universe.

| 9 |

The Role of Technology

The quest to find life beyond Earth hinges critically on our ability to observe and characterize exoplanets – planets orbiting stars other than our Sun. For decades, this challenge seemed insurmountable, but recent advancements in telescope technology have propelled us into an era of unprecedented discovery. The journey from barely detecting the presence of a distant exoplanet to analyzing the composition of its atmosphere is a testament to human ingenuity and our relentless pursuit of knowledge.

The development of increasingly sensitive detectors has been paramount. Early exoplanet detection methods, like radial velocity measurements (detecting the slight wobble of a star caused by an orbiting planet's gravity), were limited by the precision of available instruments. Modern spectrographs, however, boast vastly improved sensitivity, allowing us to detect the subtle gravitational tugs of smaller planets, including those potentially resembling Earth in size and mass. This increased sensitivity is a direct result of advancements in detector technology, encompassing both the development of more efficient photodiodes and the sophisticated algorithms used to process the collected data. These algorithms can filter out noise and isolate the faint signals indicative of planetary orbits, enhancing the signal-to-noise ratio significantly.

Beyond improved sensitivity, advancements in telescope resolution are equally crucial. Imaging exoplanets directly is extremely difficult due to the overwhelming brightness of their host stars. However, techniques like coronagraphy, which uses masks to block out starlight, have improved dramatically. Modern coronagraphs, often employing deformable mirrors to actively shape the starlight, can achieve far greater contrast, making it possible to directly image exoplanets that are much closer to their stars than was previously possible. Furthermore, advancements in adaptive optics, which compensate for the blurring effects of Earth's atmosphere, have significantly enhanced the resolution of ground-based telescopes, allowing for finer details to be resolved in images of exoplanets. The use of advanced algorithms, trained on extensive datasets, helps to further refine these images, removing remaining artifacts and enhancing the detail.

Spectral capabilities of telescopes have also experienced a transformative evolution. Spectroscopy involves analyzing the light from an exoplanet to determine its atmospheric composition. Early spectroscopic analyses were limited to detecting only the most abundant atmospheric gases. However, the advent of high-resolution spectrographs, combined with sophisticated data analysis techniques, now allows us to detect trace gases, some of which could be considered biosignatures – indicators of the presence of life. For example, the detection of oxygen, methane, or other gases in specific concentrations could potentially suggest biological activity on an exoplanet, though it is crucial to remember that abiotic processes can also produce such gases. The development of extremely large telescopes (ELTs), currently under construction, will push the boundaries of spectroscopic analysis even further, allowing us to probe the atmospheres of smaller, potentially Earth-like exoplanets with unprecedented detail.

The James Webb Space Telescope (JWST), launched in 2021, stands as a prime example of these technological leaps. Its infrared capabilities allow it to peer through dust clouds and observe exoplanets that would be invisible to optical telescopes. JWST's high sensitivity and resolution have already yielded remarkable data on exoplanet atmospheres, including the detection of water vapor and other molecules in several systems. JWST is not merely an upgrade on previous telescopes; it represents a paradigm shift in our ability to study exoplanets, a shift driven by years of incremental improvements in detector technology, optics, and data analysis techniques. The sheer amount of data collected by JWST necessitates advanced computational tools for processing and analysis, further highlighting the symbiotic relationship between technology and scientific progress.

Looking forward, even more ambitious projects are underway. The Extremely Large Telescope (ELT), a ground-based observatory with a primary mirror 39 meters in diameter, is expected to significantly surpass the capabilities of existing telescopes. Its enormous light-gathering power, coupled with advanced adaptive optics, will allow for the detection and characterization of even smaller, fainter exoplanets. Furthermore, the planned use of interferometry, which combines the light from multiple telescopes, will effectively create a telescope with a much larger aperture, significantly increasing the resolution and sensitivity.

Beyond ELTs, space-based observatories are also planned, which will overcome the limitations imposed by Earth's atmosphere. These future observatories may utilize coronagraphy and other advanced techniques to directly image and characterize exoplanets with unprecedented detail. Furthermore, the development of specialized instruments, designed to detect specific biosignatures, will be crucial. For instance, the search for signs of life might focus on specific wavelengths of light that are indicative of photosynthetic activity or other biological processes. The future of exoplanet research is not merely an extension of current capabilities, but rather a leap into new realms of exploration, enabled by the ongoing advancements in telescope technology.

The development of sophisticated data analysis techniques is just as critical as the advancements in telescope hardware. The sheer volume of data generated by these powerful telescopes requires the development of algorithms and artificial intelligence (AI) to efficiently analyze the information. AI can identify patterns, filter out noise, and prioritize data for further analysis, significantly accelerating the pace of discovery. Machine learning algorithms can be trained to identify potential biosignatures, even subtle ones that might be missed by human researchers. The integration of AI into the data analysis pipeline is not just an efficiency boost; it is a necessity given the immense volume and complexity of the data generated by modern telescopes.

The search for extraterrestrial life is a multi-faceted endeavor, requiring a holistic approach. Advances in telescope technology, while incredibly significant, are only one piece of the puzzle. The development of more sophisticated spacecraft, robotic explorers, and sample-return missions is crucial for conducting in-situ analyses of potentially habitable worlds. Furthermore, continued research into extremophiles on Earth provides vital insights into the diversity of life and the range of environments that can support it. By understanding the limits of life on Earth, we can refine our search strategies for life beyond our planet. The quest for extraterrestrial life is a journey of scientific exploration, technological innovation, and international collaboration, and advancements in telescope technology are a crucial driving force behind this exciting endeavor. The future holds immense promise, and the technological advancements described above will likely lead to discoveries that will fundamentally reshape our understanding of life in the universe.

The quest to understand the potential for life beyond Earth isn't solely reliant on our ability to observe distant planets; it also depends heavily on our capacity to physically explore them. This necessitates advancements in spacecraft design and capabilities, pushing the boundaries of what's technologically feasible. The journey from the early, rudimentary probes to the sophisticated robotic explorers of today represents a remarkable testament to human ingenuity and our relentless pursuit of knowledge.

One of the most significant hurdles in space exploration is propulsion. Early space missions relied heavily on chemical rockets, which, while powerful, are relatively inefficient in terms of fuel consumption and speed. The sheer distance to even the nearest potentially habitable exoplanets necessitates a radical improvement in propulsion systems. Several alternative propulsion technologies are under development, each offering potential advantages over traditional chemical rockets.

Ion propulsion, for instance, uses electric fields to accelerate ionized gas, providing a much higher specific impulse (a measure of fuel efficiency) than chemical rockets. While the thrust produced by ion engines is relatively low, their sustained operation over long periods allows them to achieve significantly higher velocities. This makes them ideal for long-duration missions to distant destinations. NASA's Dawn mission, which successfully orbited both Vesta and Ceres in the asteroid belt, employed ion propulsion, demonstrating its viability for deep-space exploration. Further advancements in ion propulsion, such as the development of more powerful and efficient ion sources, are crucial for future interstellar missions.

Nuclear thermal propulsion offers another promising avenue. This approach uses a nuclear reactor to heat a propellant, such as hydrogen, generating significantly higher exhaust velocities than chemical rockets. Nuclear thermal propulsion could dramatically reduce travel times to distant planets, making missions feasible that are currently impractical with chemical rockets. However, the development of nuclear thermal propulsion systems requires addressing safety concerns associated with handling radioactive materials in space. Ongoing research is focused on developing safer and more efficient reactor designs, minimizing the risks involved.

Another area of ongoing development is nuclear fusion propulsion. Fusion, the process that powers the sun, offers the potential for incredibly high specific impulse and thrust levels. While still in its early stages of development, fusion propulsion could revolutionize space travel, enabling missions to distant stars within a human lifetime. However, achieving controlled fusion on a spacecraft remains a significant technological challenge, requiring breakthroughs in materials science, plasma physics, and engineering.

Beyond propulsion, the power systems of spacecraft are crucial for their operation. Solar panels, while effective for missions within the inner solar system, become less efficient as distance from the sun increases. Radioisotope thermoelectric generators (RTGs) offer a reliable power source for long-duration missions, converting the heat generated by the decay of radioactive isotopes into electricity. RTGs have been successfully used on numerous deep-space missions, such as the Voyager probes and the Curiosity rover on Mars. However, the availability and cost of radioisotopes limit the widespread adoption of RTGs.

Advanced reactor designs could provide a more sustainable and powerful alternative to RTGs for long-duration missions. These reactors could generate significantly more power than RTGs, enabling more ambitious missions and more sophisticated scientific instrumentation. However, the development of space-based nuclear reactors requires addressing safety and regulatory concerns, ensuring the protection of both the environment and the astronauts.

The scientific instruments carried by spacecraft are equally critical. Early missions relied on relatively simple instruments, but advancements in miniaturization and sensor technology have dramatically increased the sophistication of spaceborne instruments. Modern spacecraft carry a range of sophisticated sensors, including spectrometers, cameras, radar, and laser altimeters, allowing for detailed analysis of planetary surfaces and atmospheres. These instruments provide crucial data for assessing the habitability of celestial bodies and searching for signs of life.

Miniaturization has played a pivotal role in increasing the capabilities of spacecraft. Smaller, lighter instruments allow for more payload capacity, enabling the inclusion of more sophisticated sensors and experiments. The development of micro-electromechanical systems (MEMS) has been particularly influential in this regard, enabling the creation of miniature sensors and actuators for a wide range of applications.

The development of robust and reliable robotics is fundamental for in-situ exploration. Robotic explorers can navigate challenging terrains, collect samples, and perform experiments that would be too dangerous or impractical for human astronauts. The Mars rovers, such as Curiosity and Perseverance, are prime examples of the capabilities of robotic exploration. Their sophisticated instruments allow them to analyze Martian rocks and soil, searching for signs of past or present life. Future robotic explorers will likely incorporate advanced AI capabilities, enabling them to perform more autonomous operations and adapt to unexpected situations. This will dramatically expand the scope of robotic missions, allowing them to explore more challenging and hazardous environments.

Sample-return missions represent a crucial step towards a deeper understanding of other celestial bodies. These missions involve collecting samples from a planet or moon and returning them to Earth for detailed laboratory analysis. The Apollo missions returned lunar samples that revolutionized our understanding of the moon's formation and history. Future sample-return missions to Mars or other potentially habitable bodies could provide even more significant scientific discoveries, offering potentially irrefutable evidence of past or present life beyond Earth. The technological challenges associated with sample-return missions are significant, requiring robust spacecraft designs and reliable retrieval mechanisms.

The advancements described above are interconnected and mutually reinforcing. For example, improved propulsion systems enable longer missions, necessitating more efficient power systems and robust scientific instrumentation. The development of sophisticated robotics requires advancements in AI and miniaturization, allowing for more autonomous exploration. The quest to explore and understand other celestial bodies is a complex undertaking, but the ongoing advancements in spacecraft design and capabilities are progressively increasing our reach and the depth of our investigations. As technology continues to evolve, the possibilities for exploring the cosmos and unraveling the mysteries of extraterrestrial life become increasingly real. The development of more advanced spacecraft, including those capable of interstellar travel, remains a significant long-term goal, but the progress made in recent decades offers significant cause for optimism. The future of space exploration promises to be filled with remarkable discoveries, further pushing the boundaries of our understanding of the universe and our place within it. The integration of cutting-edge materials science, improved thermal management systems, and enhanced radiation shielding will be critical in enabling these longer, more ambitious missions. The development of autonomous navigation systems and self-repairing mechanisms will also significantly enhance the robustness and resilience of future spacecraft. Ultimately, the convergence of these technological advances will pave the way for a new era of exploration, bringing us closer to answering the fundamental question of whether we are alone in the universe.

The search for extraterrestrial life is no longer confined to the realm of science fiction; it's a vibrant field of scientific inquiry driven by technological advancements that allow us to gather and analyze data on an unprecedented scale. While the development of advanced spacecraft and propulsion systems are crucial for physical exploration, the parallel advancement of data analysis and computational tools is equally vital in making sense of the vast amounts of information we collect from space. This data, ranging from telescopic observations of exoplanets to the chemical analysis of Martian soil samples, is often complex, incomplete, and noisy, necessitating sophisticated methods for interpretation.

The sheer volume of data generated by modern astronomical surveys is staggering. Telescopes like the Kepler and TESS missions have discovered thousands of exoplanets, each requiring detailed characterization. This includes determining their size, mass, orbital period, and atmospheric composition, data points that are often derived from indirect measurements requiring advanced statistical modelling. Consider, for example, the transit method, where an exoplanet's passage in front of its star causes a slight dip in the star's brightness. Accurately measuring the depth and duration of these dips requires sophisticated algorithms to filter out noise from the signal and to compensate for various confounding factors such as stellar activity. The resulting data is then further analyzed to infer properties such as the planet's radius and orbital period. This process is not merely a simple calculation; it involves complex statistical modelling to account for uncertainties and errors in the measurements.

Furthermore, spectroscopic analysis of exoplanet atmospheres presents another significant challenge. The faint signals from exoplanet atmospheres are often overwhelmed by the much brighter light from their host stars. Sophisticated algorithms are required to separate the exoplanet signal from the stellar light, a process analogous to isolating a faint whisper in a crowded room. This signal separation involves spectral deconvolution and advanced signal processing techniques to extract subtle spectral features indicating the presence of specific molecules, such as water vapor, methane, or oxygen. The detection of these biosignatures is a crucial step in assessing the potential habitability of an exoplanet and its potential to harbor life. Even then, careful statistical analysis is required to assess the significance of these detections and rule out alternative explanations.

The field of machine learning is increasingly being leveraged to accelerate the analysis of this massive dataset. Machine learning algorithms, particularly deep learning techniques like convolutional neural networks, are proving highly effective at identifying patterns and anomalies in large datasets that might be missed by human observers. These algorithms can be trained on existing datasets of known exoplanets and stars to identify potentially habitable worlds based on their observed characteristics. Such algorithms can not only sift through vast amounts of data far more efficiently than human analysts but can also identify subtle correlations and patterns that may not be immediately apparent. For example, a machine learning model could potentially identify subtle spectral features indicative of life that might be missed by human analysis, accelerating the pace of discovery.

Beyond exoplanet characterization, computational tools are equally critical in analyzing data from robotic missions. Data from Mars rovers like Curiosity and Perseverance, including images, spectrometer readings, and chemical analyses of Martian rocks and soil, requires sophisticated image processing and data fusion techniques to interpret. For example, identifying potential fossilized microbial life requires image analysis to discern microstructures from geological formations, a task that can be aided by machine learning algorithms trained to recognize biological features. Similarly, the analysis of chemical data requires sophisticated statistical methods to differentiate between the chemical signatures of abiotic processes (those not involving life) and potential biosignatures. Computational modelling is also used to simulate the conditions on early Mars and assess the potential for the evolution of life under those conditions.

The development of new algorithms specifically tailored to astrobiological questions is an active area of research. This includes algorithms that can: (1) efficiently process and analyze the massive datasets generated by future space missions; (2) identify subtle biosignatures in exoplanet atmospheres and planetary surfaces; (3) build detailed models of the evolution of planetary systems and their potential habitability; and (4) integrate data from multiple sources to build a comprehensive understanding of extraterrestrial environments.

Furthermore, the development of robust statistical methods is crucial for interpreting the results of these analyses and assessing the confidence in our conclusions. The detection of biosignatures, for example, requires careful statistical analysis to differentiate between true signals and random fluctuations in the data. False positives are a significant concern in astrobiological research, and rigorous statistical methods are essential to ensure the reliability of our findings. This involves not only the application of established statistical techniques, but also the development of new methods tailored to the unique challenges of astrobiological data analysis. This is particularly important given the potential for bias in data interpretation and the inherent difficulty of definitively proving or disproving the existence of extraterrestrial life.

The role of open-source software and data sharing is also critical. Making data and software readily available to the wider scientific community fosters collaboration and accelerates progress. This allows researchers to build upon each other's work, verify results, and develop more powerful analytical tools. The development of standardized data formats and analysis pipelines is crucial for ensuring interoperability and facilitating data exchange between different research groups. This collaborative approach is vital given the complexity and interdisciplinarity of astrobiological research, bringing together expertise from astronomy, biology, geology, chemistry, and computer science.

In conclusion, the quest for extraterrestrial life depends not only on our ability to explore the cosmos but also on our capacity to analyze the immense volume of data generated by these explorations. The development of sophisticated computational tools and data analysis techniques, including advanced algorithms and machine learning, is crucial for making sense of this information, identifying potential biosignatures, and ultimately answering one of the most fundamental questions in science: Are we alone? The ongoing development and refinement of these tools promise to dramatically accelerate our understanding of the universe and our place within it, ushering in a new era of discovery in astrobiology. The integration of these advanced computational techniques with new space exploration missions will prove instrumental in achieving a deeper, more comprehensive understanding of the potential for life beyond Earth. The exciting prospect is that we are on the cusp of a new era of discovery, with the technology to process and interpret the vast amounts of information necessary to potentially answer this profound question. As our technological capabilities continue to improve, the likelihood of uncovering evidence of life beyond Earth, whether past or present, increases exponentially, fueled by the innovative analytical tools shaping the future of astrobiology.

The current technological landscape, while impressive, presents significant limitations in our quest to find extraterrestrial life. Our telescopes, while powerful, are still constrained by the vast distances involved. Even the most advanced instruments struggle to detect subtle biosignatures in the atmospheres of exoplanets, often drowned out by the overwhelming brightness of their host stars. Improved sensitivity and resolution are critical, requiring advancements in both optical and spectroscopic technologies. This means developing instruments that can effectively filter out stellar noise and isolate the faint signals from exoplanets, allowing for the detection of trace gases that could indicate the presence of life. Furthermore, the development of advanced adaptive optics systems, which compensate for the blurring effects of Earth's atmosphere, will be vital in enhancing the resolution of ground-based telescopes, allowing us to probe even fainter exoplanet atmospheres with greater detail. Space-based telescopes, operating free from atmospheric interference, are also crucial. Future missions should prioritize the development of larger aperture telescopes equipped with more advanced spectrometers and coronagraphs – instruments designed to block out the light from a star, enabling direct imaging of exoplanets and the detailed analysis of their atmospheres. The development of mid-infrared telescopes, capable of detecting the heat signatures of exoplanets, will also be invaluable in the search for habitable worlds. These advancements will not only improve our ability to detect exoplanets but also enhance our capability to study their atmospheric composition with unprecedented precision.

Beyond detection, the characterization of exoplanet atmospheres remains a considerable hurdle. Current techniques rely on indirect measurements, such as transit spectroscopy, which only provide limited information. Direct imaging and spectroscopy of exoplanet atmospheres, while challenging, are crucial for a more complete understanding of their potential habitability. This requires the development of extremely high-contrast imaging techniques that can effectively suppress the light from the host star, allowing for detailed spectroscopic analysis of the exoplanet's atmosphere. Advancements in interferometry, a technique that combines the light from multiple telescopes to achieve higher resolution, will play a vital role in this endeavor. This would allow astronomers to achieve resolution comparable to or exceeding that of the Hubble Space Telescope. Moreover, the development of advanced spectroscopic instruments capable of detecting a wider range of wavelengths and with enhanced sensitivity will be critical in identifying the subtle spectral features that could indicate the presence of biosignatures. This includes the development of instruments capable of detecting the signatures of complex organic molecules, often considered as strong indicators of life, and the development of sensitive detectors capable of distinguishing between biotic and abiotic processes.

Propulsion systems represent another major technological bottleneck. The vast distances between stars make interstellar travel a daunting challenge. Current propulsion technologies, relying primarily on chemical rockets, are far too slow for interstellar missions. The development of advanced propulsion systems, such as fusion propulsion or ion propulsion, is vital for significantly reducing travel times and enabling more ambitious exploration missions. Fusion propulsion, for example, offers the potential for orders of magnitude increase in speed compared to chemical rockets. However, this technology is still under development, and significant engineering breakthroughs are required before it can be applied to interstellar travel. Similarly, advancements in ion propulsion, which uses electric fields to accelerate ions, could enable longer duration missions with significantly reduced fuel consumption. These advanced propulsion systems would allow us to reach nearby star systems within a reasonable timeframe, enabling the deployment of robotic probes and eventually, perhaps, crewed missions. The development of robust and reliable spacecraft capable of withstanding the harsh conditions of interstellar space is also critical. This includes developing shielding against cosmic radiation, micrometeoroids, and other hazards. Furthermore, long-duration missions will require autonomous systems capable of handling unexpected situations and conducting scientific investigations with minimal human intervention.

Data analysis and computational power also need considerable improvement. The sheer volume of data generated by future missions will be astronomical, requiring sophisticated algorithms and high-performance computing capabilities to process and analyze. The development of advanced machine learning algorithms capable of identifying subtle patterns and anomalies in large datasets will be crucial. These algorithms can be trained on existing data to identify potential biosignatures, analyze images, and interpret spectroscopic data far more efficiently than human analysts. Further, this includes the development of new algorithms that can handle incomplete or noisy data, which is often the case in astrobiological research. Advances in high-performance computing, including the development of quantum computers, have the potential to revolutionize data analysis in astrobiology. These powerful computers can handle complex calculations far beyond the capabilities of classical computers, allowing for more detailed simulations of planetary environments and the evolution of life. Coupled with the advanced data analysis methods, quantum computers could potentially unlock hidden patterns in large datasets that traditional methods fail to detect. The development of robust statistical methods is also essential for interpreting the results of data analysis and ensuring the reliability of our findings. This requires addressing challenges such as biases in data interpretation and carefully considering the possibility of false positives.

Finally, international collaboration is critical. The search for extraterrestrial life is a global endeavor that requires the combined expertise and resources of many nations. International collaborations, such as the ongoing efforts within the SETI community and the cooperation between space agencies, are crucial for sharing data, coordinating missions, and developing new technologies. Open-source software and data sharing are also critical for accelerating progress. Making data and software readily available to the wider scientific community fosters collaboration, allows for the verification of results, and enables researchers to build upon each other's work. The development of standardized data formats and analysis pipelines is crucial for ensuring interoperability and facilitating data exchange between different research groups, enabling a more effective global response to this complex scientific challenge.

In summary, the search for extraterrestrial life requires a significant leap forward in our technological capabilities. More sensitive instruments, more powerful telescopes, more efficient propulsion systems, and advanced data analysis techniques are all crucial for achieving this ambitious goal. However, this is not merely a matter of technological advancement; the quest to answer the question of our cosmic solitude necessitates a holistic approach, integrating the best scientific minds, combining resources, and forging international collaborations. The future of astrobiology relies on bridging the gap between our current technological limitations and the vast technological needs that the exploration of this compelling field demands. The investment in these advancements will not only advance our understanding of the universe but potentially revolutionize our place within it. The combination of cutting-edge technology with rigorous scientific methodology promises a new era of discovery, one that will redefine our understanding of life itself and our place in the cosmos. The technological hurdles are immense, yet the potential rewards – a fundamental shift in our understanding of life, the universe, and everything – are beyond measure.

The previous sections highlighted the technological hurdles facing the search for extraterrestrial life, from the limitations of current telescopes to the challenges of interstellar travel and data analysis. However, overcoming these challenges isn't solely a matter of incremental technological improvements; it demands a fundamental shift in our approach. The search for life beyond Earth is inherently interdisciplinary, requiring a synergistic collaboration between experts from seemingly disparate fields. The success of this quest hinges not on individual breakthroughs but on the collective power of a unified, collaborative effort.

The development of advanced telescopes, for instance, requires a close working relationship between optical engineers, astronomers, and physicists. Optical engineers design and build the intricate systems of lenses, mirrors, and detectors that form the heart of a telescope. Astronomers, with their expertise in celestial mechanics and astrophysics, guide the design process, ensuring that the instrument is optimized for the specific challenges of observing exoplanets and detecting biosignatures. Physicists, particularly those specializing in materials science and quantum optics, contribute to the development of advanced detector technologies capable of capturing even the faintest signals from distant worlds. This multifaceted collaboration ensures that the resulting instrument is not only technologically feasible but also scientifically effective.

The development of advanced propulsion systems similarly necessitates a convergence of diverse expertise. Engineers specializing in aerospace engineering, plasma physics, and materials science must work together to design, build, and test innovative propulsion technologies like fusion rockets or advanced ion thrusters. The challenges are immense, ranging from the containment of superheated plasma in fusion reactors to the development of lightweight, durable materials capable of withstanding the rigors of interstellar travel. The success of such endeavors relies not just on the ingenuity of individual researchers but on the coordinated efforts of diverse teams tackling different aspects of the problem, each contributing their unique skills and perspectives.

The analysis of the vast datasets generated by future missions presents another area where interdisciplinary collaboration is paramount. Astronomers and astrobiologists provide the scientific context, defining the key parameters and biosignatures to search for. Computer scientists develop the sophisticated algorithms and machine learning models capable of analyzing terabytes, petabytes, or even exabytes of data, identifying patterns and anomalies that might indicate the presence of life. Statisticians and mathematicians ensure the rigor and reliability of the data analysis, minimizing biases and validating results. This intricate interplay of expertise is crucial for effectively extracting meaningful information from the deluge of data produced by advanced telescopes and space probes.

Furthermore, the design and implementation of robotic missions to other planets and moons requires a team effort that blends expertise from many areas. Robotics engineers create the autonomous systems capable of navigating complex terrain, collecting samples, and conducting experiments in challenging environments. Geologists and planetary scientists plan the scientific objectives, selecting the most promising locations for exploration. Biochemists and molecular biologists devise experiments to search for signs of past or present life. Software engineers write the intricate codes that govern the robots' actions, ensuring they can function autonomously and adapt to unexpected situations. This collaborative undertaking is essential for maximizing the scientific yield of these expensive and complex missions.

Beyond the strictly technical aspects, social scientists and ethicists play a vital role. The potential discovery of extraterrestrial life would have profound implications for humanity, impacting our philosophical understanding of our place in the universe and raising ethical dilemmas regarding contact and exploration. Social scientists can help anticipate and manage the social and psychological consequences of such a discovery. Ethicists provide frameworks for responsible conduct, helping to shape international policies and guidelines for the exploration and interaction with extraterrestrial life. Their input is crucial in navigating the complex ethical questions that will inevitably arise.

The benefits of this interdisciplinary approach are manifold. First, it fosters innovation. By bringing together experts from diverse backgrounds, it creates an environment where novel ideas and solutions can emerge from the cross-pollination of knowledge and perspectives. Second, it accelerates the pace of discovery. By streamlining the process of technology development and data analysis, collaborative efforts can lead to breakthroughs that would be unattainable through isolated research efforts. Third, it enhances the reliability and validity of scientific findings. Collaboration promotes peer review, rigorous testing, and the critical evaluation of results, reducing the risk of errors and biases. Finally, international collaboration is essential to maximize the impact of this global endeavor. Sharing resources, expertise, and data across national boundaries ensures that the search for extraterrestrial life is a unified and coordinated effort, maximizing the chances of success.

In conclusion, the search for extraterrestrial life is not simply a technological challenge; it's a societal imperative, requiring a coordinated, interdisciplinary effort. It demands a fundamental shift from isolated, discipline-specific research towards a collaborative paradigm that integrates the diverse expertise needed to address this grand scientific quest. The success of this endeavor hinges not on individual breakthroughs but on the synergistic power of a global network of scientists, engineers, and technologists working together towards a common goal. The potential rewards—a deeper understanding of life, the universe, and our place within it—justify the investment in this collaborative, interdisciplinary approach. The collaborative spirit, characterized by open communication, data sharing, and mutual respect, will ultimately determine our success in uncovering the secrets of the cosmos and answering one of humanity's most profound questions: Are we alone? The answer, it seems, lies not only in technological advancement but also in the strength of our collective effort. The coming decades will be a testament to the power of interdisciplinary collaboration, demonstrating that humanity's greatest discoveries are often achieved through concerted, multifaceted endeavors.

| 10 |

Philosophical Implications

The previous sections outlined the immense technological and collaborative challenges inherent in the search for extraterrestrial life. However, the potential discovery of even microbial life beyond Earth carries implications far exceeding the purely scientific. It would trigger a profound philosophical reassessment of humanity's place and significance in the universe, a reassessment that could reshape our self-understanding in fundamental ways. The Copernican Revolution, which dethroned Earth from its geocentric position, was a seismic shift in human understanding. The discovery of extraterrestrial life would be a comparable, if not more profound, upheaval, forcing us to confront our anthropocentric biases and reconsider our unique status in the cosmos.

For centuries, humanity has occupied a privileged position in our worldview. Religious and philosophical traditions often placed humankind at the center of creation, endowed with a unique spirit or purpose. Even secular scientific thought, until relatively recently, implicitly assumed a certain exceptionalism for human life. The sheer scale of the universe, with its billions of galaxies and trillions of stars, already challenges this anthropocentric view. But the discovery of extraterrestrial life would not merely challenge it; it would shatter it. The existence of life elsewhere would immediately demonstrate that the emergence of life, even intelligent life, is not a unique or improbable event. It would transform the question of "Are we alone?" from a scientific query to a question answered definitively in the negative.

This shift in perspective would have far-reaching consequences for our self-perception. If life is not unique to Earth, what does it mean to be human? If other intelligent species exist, do they possess similar levels of consciousness, morality, and technology? Or are their experiences, values, and societal structures radically different from our own? The potential diversity of extraterrestrial life forms could expand our understanding of what it means to be alive, challenging our very definition of life itself. We might discover life forms with entirely different biochemistries, radically altering our assumptions about the fundamental requirements for life. Or we might encounter intelligences far surpassing our own, forcing us to reconsider our self-proclaimed intellectual superiority.

The impact on religious and spiritual beliefs would be equally significant. Many religions place humanity at the center of a divine plan, granting us a special relationship with the creator. The discovery of extraterrestrial life would directly challenge these anthropocentric theological views, forcing a reevaluation of religious doctrines and beliefs. The existence of other intelligent beings would raise questions about the nature of God, the purpose of creation, and the unique status of humanity within a divine scheme. Some might find solace and renewed meaning in the discovery of a larger cosmic community, while others might grapple with the implications for their faith, leading to profound theological debates and the restructuring of religious belief systems.

Beyond the religious and spiritual spheres, the discovery of extraterrestrial life would have profound implications for our ethical and political systems. Questions of contact, exploration, and interaction with extraterrestrial civilizations would demand careful consideration. International cooperation would be paramount, as any encounter would necessitate a unified, global approach to prevent conflicts and ensure responsible behavior. Ethical guidelines and international treaties would be essential to establish frameworks for peaceful interaction and the protection of both terrestrial and extraterrestrial life. The potential for exploitation, conflict, or the transmission of terrestrial diseases would necessitate thoughtful planning and regulation.

Moreover, the discovery of extraterrestrial life could drastically alter our societal priorities. The scale of the challenge posed by the search for extraterrestrial life—a quest demanding international cooperation, massive scientific investments, and potentially paradigm shifts in our understanding of life and the universe—could force a re-evaluation of our current societal values. The sheer magnitude of such a discovery could unite humanity in a common purpose, transcending national borders and ideological differences. The focus might shift from narrow, national interests to a broader, cosmic perspective, fostering international cooperation and a shared sense of destiny.

Conversely, the discovery could also exacerbate existing societal divisions. If extraterrestrial life is discovered to be more advanced than humanity, it could fuel feelings of inadequacy, insecurity, or even fear. Competition for resources and dominance could escalate, potentially resulting in conflict. The implications for existing power structures and societal hierarchies would need to be carefully considered, as the newfound cosmic perspective might reshape our political landscapes and challenge existing power dynamics. The economic implications are also substantial. The development of advanced technologies necessary for interstellar communication and travel would represent a massive economic endeavor, potentially reshaping global economies and investment priorities.

Furthermore, the discovery could impact our understanding of history and our place within the timeline of the universe. Our current understanding of history is largely anthropocentric, focusing on human events and achievements. The discovery of extraterrestrial life, especially intelligent life with a long history, would necessitate a broader, more inclusive perspective. Human history would become just one chapter in a much larger cosmic narrative, contextualized within the broader timeline of life in the universe. This perspective could foster humility, challenging our assumptions about the significance of human history and our place within the grand scheme of the universe.

Ultimately, the discovery of extraterrestrial life would necessitate a profound reevaluation of our place in the universe. It would force us to confront our anthropocentric biases, redefine our understanding of life, and reconsider our ethical and societal frameworks. The impact would be far-reaching and transformative, impacting every aspect of human life and thought. The coming decades may well witness a paradigm shift, not only in our scientific understanding of the cosmos but also in our fundamental understanding of ourselves and our place within the vast expanse of space and time. The search for extraterrestrial life, therefore, is not merely a scientific endeavor; it's a philosophical journey into the heart of our own existence. It is a quest to understand not just the universe around us, but the meaning and purpose of our lives within it. The outcome, however monumental, would inevitably rewrite the story of humanity.

The potential discovery of extraterrestrial life would undoubtedly trigger a profound reassessment of religious and spiritual beliefs across the globe. For centuries, many religious traditions have presented humanity as central to a divine plan, often uniquely created in God's image and given dominion over the Earth. The sheer existence of another intelligent species, let alone multiple species, would directly challenge this anthropocentric view of creation.

The question of whether these extraterrestrial beings are also imbued with a soul, a spirit, or some form of divine connection would ignite extensive theological debates. Some might argue that these beings are part of a larger cosmic plan, expanding the scope of divine creation and providing a broader context for humanity's role within it. Others might view extraterrestrial life as evidence against a creator, challenging existing notions of divine intervention and the uniqueness of humanity.

Consider the Abrahamic faiths, for example. The creation narratives in the Bible and the Quran depict humanity as uniquely created, given stewardship over the Earth, and often positioned as the pinnacle of God's creation. The discovery of extraterrestrial intelligence could force a reevaluation of these narratives, prompting a deeper examination of the meaning of "image of God" and the implications of a potentially diverse cosmic community created by the same divine hand. Would the existence of alien civilizations require a reinterpretation of divine grace and salvation? Would it necessitate a broadening of theological concepts such as the covenant, chosen people, or the very definition of sin? These questions would deeply challenge fundamental tenets of many belief systems, prompting intense theological debates and potentially leading to revisions in religious doctrines.

Similarly, many Eastern religions, such as Hinduism and Buddhism, which often focus on the interconnectedness of all beings and the cyclical nature of existence, might find the discovery of extraterrestrial life less disruptive. The concept of reincarnation, for example, could be easily extended to include extraterrestrial life forms, integrating them into the broader cosmic cycle of birth, death, and rebirth. However, even within these frameworks, questions regarding the karma and dharma of extraterrestrial beings, and their potential interaction with human beings, would necessitate theological reinterpretation. Moreover, the potential for vastly different forms of consciousness and spiritual practices among extraterrestrials could further enrich the understanding of spiritual evolution and universal principles.

Beyond organized religions, the impact on personal spirituality could be equally profound. For many, the belief in a higher power is linked to the perceived uniqueness and significance of human life. If life is not unique to Earth, this sense of uniqueness might be challenged, potentially leading to a reevaluation of personal beliefs and spiritual practices. Some might experience a sense of existential crisis, questioning the meaning and purpose of their lives in a vastly larger, more complex universe. Others, however, might find a renewed sense of wonder and awe, perceiving extraterrestrial life as further evidence of a profound and mysterious cosmos, enriching rather than diminishing their spiritual understanding.

The interaction between science and faith is also central to this discussion. Historically, the relationship between science and religion has been characterized by periods of conflict and cooperation. The discovery of extraterrestrial life could, arguably, represent a moment of convergence, where scientific discovery challenges and simultaneously deepens our spiritual understanding. Science would provide the evidence of extraterrestrial life, while religious and spiritual perspectives would help us interpret the implications of that discovery, providing frameworks for understanding our place in a broader cosmic context.

It's important to avoid a simplistic dichotomy between science and faith. The scientific method relies on empirical evidence, while faith often involves belief beyond empirical verification. However, both science and faith seek to understand the universe and our place within it. The discovery of extraterrestrial life could thus become a catalyst for a renewed dialogue between science and religion, fostering a more nuanced understanding of the relationship between these two domains of human inquiry. This dialogue could help to bridge the gap between empirical observation and spiritual interpretation, leading to a more comprehensive and integrated understanding of reality.

Furthermore, the discovery of extraterrestrial life could spark a renewed interest in philosophical questions about the nature of consciousness, the meaning of life, and the purpose of existence. These questions have been central to philosophical inquiry for millennia, and the existence of extraterrestrial life would provide new data points to inform these discussions. Consider the question of consciousness, for example. If we were to discover life forms with radically different forms of consciousness, this could challenge our very understanding of what consciousness is and how it emerges. Similarly, the discovery of extraterrestrial life could shed new light on the question of the meaning of life, challenging our anthropocentric biases and prompting us to consider the multitude of potential meanings and purposes that life might encompass.

The ethical implications of contact with extraterrestrial civilizations also have profound implications for religious and spiritual beliefs. The potential for exploitation, conflict, or even the spread of terrestrial diseases would raise serious ethical dilemmas that require careful consideration informed by religious and spiritual values. Questions about the moral responsibilities toward other intelligent beings, the potential for interspecies cooperation or conflict, and the implications of different value systems and ethical frameworks would demand critical examination. These considerations might lead to a re-evaluation of traditional moral codes and the development of new ethical frameworks capable of addressing the challenges and opportunities of interstellar interaction. The ethical dimension, therefore, becomes an integral aspect of the philosophical and religious implications of extraterrestrial contact, demanding a holistic approach that balances scientific understanding with ethical and spiritual considerations.

In conclusion, the potential discovery of extraterrestrial life would not only reshape our scientific understanding of the universe but would also profoundly alter our religious and spiritual beliefs, creating a new paradigm shift in human thought and experience. It would challenge existing theological frameworks, encourage a reassessment of our place within the cosmos, and necessitate a renewed dialogue between science and faith. The outcome will undoubtedly be complex, multifaceted, and deeply personal, varying across individuals and cultural contexts, but it is certain to be one of the most significant events in human history, forcing us to confront fundamental questions about our existence and the nature of reality itself. The impact will likely be felt not only in our understanding of the universe but also in the very fabric of human society and our individual spiritual journeys. The resulting shift in perspective could be profoundly transformative, leading to a richer, more inclusive, and ultimately, more meaningful understanding of life itself.

The prospect of extraterrestrial life forces a confrontation with some of humanity's most fundamental existential questions. For millennia, philosophers and theologians have grappled with the meaning of existence, the purpose of life, and the uniqueness of our species. The discovery of even a single extraterrestrial civilization, let alone a multitude of them, would irrevocably alter the parameters of this age-old debate. The anthropocentric view, which places humanity at the center of existence, would be profoundly challenged, requiring a radical re-evaluation of our place in the cosmic scheme.

One of the most immediate and significant challenges would be to our understanding of human uniqueness. Many belief systems, both religious and secular, implicitly or explicitly posit humans as special, even unique, within the universe. This sense of uniqueness often informs our sense of self-worth, our moral systems, and our understanding of our purpose. If other intelligent life forms exist, and particularly if they exhibit advanced technology or cultural sophistication, our presumed uniqueness would crumble. This doesn't necessarily imply a devaluation of human life, but it does necessitate a significant recalibration of our self-perception. We would need to redefine what it means to be human in a universe that is demonstrably teeming with life. Instead of being the sole intelligent species, we would be one among many, part of a larger cosmic community, each with its unique characteristics, evolutionary pathways, and cultural expressions.

This shift in perspective could trigger a profound existential crisis for some. The realization that we are not alone, that our existence is not exceptional, could lead to feelings of insignificance or meaninglessness. The familiar narratives that provide comfort and meaning might suddenly seem inadequate, leaving a void that needs to be filled with a new understanding of our place in the universe. The search for meaning, therefore, would become even more critical in a post-discovery world.

However, the discovery of extraterrestrial life could also lead to a profound sense of wonder and awe, a realization of the vastness and complexity of the cosmos. The existence of other intelligent life forms could spark a renewed appreciation for the fragility and preciousness of life itself. It could encourage a deeper exploration of our own existence, prompting us to cherish our unique experiences and contributions to the cosmic tapestry. Instead of feeling diminished, we might experience a profound expansion of our consciousness, a greater understanding of the interconnectedness of all things, and a deeper appreciation for the inherent value of life, irrespective of its origin or location.

The question of the meaning of life itself would also be significantly impacted. If our existence is not unique, then what is our purpose? Are we merely a footnote in the grand cosmic narrative, or do we have a significant role to play? The answers to these questions are unlikely to be simple or universal. Different individuals and cultures will likely arrive at different conclusions, based on their own values, beliefs, and perspectives. However, the very act of grappling with these questions in the context of a multi-species universe will inevitably lead to a richer and more nuanced understanding of the meaning of human existence.

The discovery of extraterrestrial life might also force a reconsideration of our ethical frameworks. Currently, our ethical systems are largely anthropocentric, focused on the well-being of human beings. If we encounter other intelligent species, we will need to develop new ethical frameworks that encompass a broader range of beings and interests. This will require careful consideration of the potential for conflict, cooperation, and exploitation in interstellar interactions. It will also require us to develop new standards for interspecies communication, diplomacy, and coexistence. The development of such ethical frameworks would be a monumental challenge, but also a tremendous opportunity to expand our moral understanding and create a more just and equitable universe.

Furthermore, the existence of extraterrestrial life could have profound implications for our understanding of consciousness. If we discover life forms with radically different cognitive abilities, communication styles, or even forms of consciousness, it could challenge our very definition of intelligence and sentience. We might need to reevaluate our assumptions about the nature of consciousness, recognizing the possibility of a vast spectrum of cognitive experiences and capacities beyond our current comprehension. This could lead to a deeper appreciation of the diversity of life and consciousness in the universe and could potentially revolutionize fields like neuroscience, psychology, and philosophy of mind.

The implications extend beyond the purely intellectual. The discovery of extraterrestrial life could significantly affect our societal structures, political systems, and economic realities. A sudden influx of knowledge and technology from an advanced civilization could disrupt our existing power structures, leading to political instability or potentially even the collapse of existing social orders. Economically, the discovery could lead to massive wealth redistribution, the creation of new industries, and potentially the resolution of resource scarcity issues. However, it could also lead to economic inequality and exploitation, exacerbating existing social problems. The social and political ramifications, therefore, are as significant as the scientific and philosophical ones, necessitating careful planning and proactive adaptation.

Finally, the search for extraterrestrial life and the potential discovery thereof represent a powerful catalyst for global unity. The common goal of understanding our place in the universe could transcend national borders, cultural differences, and ideological divides. This shared endeavor could create a sense of collective identity and purpose, potentially fostering international cooperation and collaborative efforts on an unprecedented scale. It could create a shared narrative that unites humanity and redirects our collective energies towards a shared future, helping us overcome the divisive forces that plague our planet. While challenges and conflicts are inevitable, the overarching goal of understanding our place in a potentially vast and diverse universe could act as a unifying force, propelling humanity towards a more peaceful and collaborative future. This unifying purpose could be transformative, shifting focus from earthly conflicts toward a shared quest for understanding the mysteries of the cosmos. The pursuit of knowledge about extraterrestrial life, therefore, represents not only a scientific endeavor but also a potent force for potential global harmony. The quest itself could bring humanity closer together, creating a shared narrative and a common goal that transcends our differences.

The discovery of extraterrestrial life, particularly intelligent life, would profoundly reshape our understanding of consciousness and intelligence. For millennia, humanity has grappled with defining these concepts, largely within the confines of our own limited experience. Our definitions, inherently anthropocentric, are built upon observations of human cognition, behavior, and self-awareness. Encountering alien intelligence would necessitate a radical expansion of these definitions, forcing us to confront the possibility of forms of consciousness and intelligence vastly different from our own.

Consider the very nature of intelligence. We tend to equate intelligence with problem-solving abilities, abstract reasoning, and the capacity for complex communication. However, these are characteristics shaped by our evolutionary history and environmental pressures. An extraterrestrial species might exhibit intelligence in entirely different ways, perhaps manifesting in forms we currently fail to recognize as intelligent. Their cognitive processes could be based on fundamentally different principles, utilizing different sensory modalities or employing computational methods beyond our current understanding. Imagine a silicon-based life form whose "thinking" is based on quantum entanglement, operating on scales and speeds far exceeding our biological limitations. Their "intelligence" might appear chaotic or incomprehensible to us, yet perfectly rational and efficient from their perspective. Or, consider a species whose communication relies not on language as we understand it, but on subtle shifts in bioluminescence or complex pheromonal signals. Their intellectual achievements might be encoded in patterns we can't readily perceive, requiring innovative technological solutions to decipher their sophisticated communication.

This leads to the question of sentience. We, as humans, possess a subjective experience of consciousness – a sense of self, a capacity for feeling, and an awareness of our own existence. We assume that sentience is inextricably linked to intelligence, but is this a valid assumption? An extraterrestrial species might possess high levels of intelligence, capable of complex problem-solving and technological advancement, yet lack the subjective experience of consciousness we associate with sentience. Their intelligence might be purely functional, devoid of the emotional and experiential richness of human consciousness. Conversely, a species might exhibit low levels of what we consider "intelligence" – lacking advanced technology or complex communication – yet possess a profound and nuanced inner life, a rich tapestry of subjective experience far beyond our current understanding.

Furthermore, the concept of "intelligence" itself may be fundamentally flawed when applied across vastly different species. Our current methods for measuring intelligence, typically based on human-centric metrics like IQ scores, are inherently biased and inadequate for evaluating alien cognition. These tests are designed to measure specific types of human intelligence, primarily verbal and logical reasoning. An extraterrestrial species might excel in other areas, exhibiting remarkable skills in spatial reasoning, intuitive understanding, or sensory perception, abilities that our current tests fail to capture. They might be masters of environmental adaptation, possessing intricate biological systems that allow them to thrive in extreme conditions, demonstrating a form of "intelligence" deeply embedded in their physiology. Their intelligence might be expressed not through abstract thought, but through a seamless integration with their environment, a harmonious coexistence that transcends our narrow definition of intellect.

The differences between human intelligence and potential extraterrestrial types could stem from variations in evolutionary pressures, environmental contexts, and the fundamental nature of their biological or technological makeup. On Earth, the diversity of life forms reflects the remarkable adaptability of life to vastly different environments. From the microscopic complexity of bacteria to the intricate social structures of primates, intelligence has manifested in countless forms. It's not unreasonable to expect the same – or even greater – diversity in the cosmos. Extraterrestrial intelligence might have evolved in environments radically different from Earth, shaping their cognitive abilities and sensory systems in ways we cannot easily comprehend. A species that evolved on a planet with extreme gravitational forces might possess a physical and cognitive structure fundamentally different from ours. Their experience of reality might be profoundly altered by their environmental context, shaping their intelligence and consciousness in unexpected ways. For example, a species evolving in a low-light environment could possess highly sensitive infrared vision, leading to a different form of intelligence built upon this unique perceptual capacity. Their cognitive maps of the world might be entirely different from ours, reflecting their unique sensory experiences.

The potential for radically different forms of communication presents another significant challenge. Human language is a complex system of symbols and grammar that has evolved over millennia. It's deeply intertwined with our social structures and cognitive processes. Extraterrestrial communication might employ entirely different modalities, such as telepathy, bioluminescence, or even manipulation of gravitational fields. Decoding such communication would require a paradigm shift in our understanding of information transfer, potentially involving technological breakthroughs in sensory perception and data analysis. We might need to develop entirely new methods for interpreting complex patterns and identifying meaningful signals, pushing the boundaries of our technological and cognitive capacities. Imagine a species communicating through subtle alterations in the electromagnetic field around them, information encoded in patterns that are imperceptible to our current senses. The effort to decode this form of communication could require decades of research and the development of new technologies for detecting and interpreting such subtle signals.

Beyond the challenges, the discovery of extraterrestrial life and their forms of intelligence offers an unparalleled opportunity for scientific advancement. Our understanding of consciousness, neuroscience, and cognitive science would be revolutionized by the study of alien minds. We could gain insights into the fundamental principles of consciousness, exploring the biological and physical underpinnings of subjective experience. This interspecies comparison would allow us to test existing theories of consciousness and potentially formulate new ones. The diversity of cognitive architectures might reveal fundamental principles governing intelligence, revealing commonalities and differences in how consciousness can emerge in vastly different biological or technological systems. This would not only redefine our understanding of intelligence but could potentially lead to breakthroughs in fields like artificial intelligence, inspiring new algorithms and computational models based on alien cognitive processes.

The ethical implications of encountering extraterrestrial life with differing forms of intelligence are equally profound. How do we ensure ethical treatment and respect for alien life forms whose consciousness and sentience are radically different from our own? How do we navigate interspecies relations, balancing potential benefits with risks of exploitation or conflict? Developing a universally applicable ethical framework for interactions with extraterrestrial life requires a paradigm shift in our moral reasoning, expanding our empathy and understanding beyond the boundaries of our own species. It requires meticulous consideration of cultural, biological, and technological differences, acknowledging the inherent uncertainties and the potential for unexpected challenges. The development of these frameworks will require intensive interdisciplinary collaboration involving scientists, ethicists, philosophers, and policymakers. The potential impact of such a discovery demands an informed and proactive approach to ensure the well-being of all involved – both human and extraterrestrial.

In conclusion, the prospect of extraterrestrial life, particularly intelligent life, forces us to confront the limitations of our current understanding of consciousness and intelligence. It necessitates a radical re-evaluation of our anthropocentric perspectives and a willingness to embrace the possibility of forms of consciousness and intelligence beyond our current comprehension. The challenges are immense, yet the potential rewards – scientific breakthroughs, ethical advancements, and a profound expansion of our understanding of the universe and our place within it – are equally immense. The journey towards understanding the nature of consciousness and intelligence in a universe teeming with life is a journey of profound discovery, one that promises to reshape our understanding of ourselves and our place in the cosmos in ways we can only begin to imagine.

The discovery of extraterrestrial life, particularly intelligent life, would irrevocably alter the trajectory of humanity's future. The impact would be far-reaching, affecting not only our scientific understanding of the universe but also our social structures, political systems, and even our fundamental sense of self. Several scenarios are plausible, each with its own unique set of challenges and opportunities.

One potential outcome is a period of intense scientific collaboration and technological advancement. The knowledge gained from studying extraterrestrial biology, technology, and culture would be immense, potentially accelerating technological progress at an unprecedented rate. We might witness breakthroughs in medicine, materials science, energy production, and space travel, leading to a new era of prosperity and human flourishing. Imagine gaining access to technologies that would revolutionize energy production, eliminating our reliance on fossil fuels and ushering in an era of sustainable and abundant energy. Or consider advancements in medical technology, leading to cures for diseases that currently plague humanity. The potential benefits are staggering, potentially transforming our civilization in ways we can only currently dream of.

However, this positive scenario is not without its potential downsides. The influx of new information and technologies could overwhelm our existing social and political structures, leading to instability and conflict. Competition for resources, both terrestrial and extraterrestrial, could escalate, leading to a new arms race, possibly even interstellar conflict. The management of such advanced technologies could prove challenging, posing risks of unforeseen consequences. The unequal distribution of benefits could exacerbate existing societal inequalities, creating further divisions and tensions. The potential for misuse of advanced technologies, particularly in the realm of weaponry, is a sobering consideration. This requires careful planning and international cooperation to avoid a dystopian future where advanced technologies are used for destructive purposes.

Another possible scenario involves a significant shift in our philosophical and religious beliefs. The discovery of extraterrestrial life would challenge existing religious doctrines and cosmological models, leading to a reassessment of humanity's place in the universe. This could lead to a period of spiritual and intellectual upheaval, but it could also foster a new sense of unity and shared purpose, bringing together people from diverse backgrounds and beliefs under a shared sense of wonder and awe. The paradigm shift might be gradual, with existing beliefs evolving to incorporate new knowledge, or it could be more dramatic, with the adoption of new belief systems reflecting our newly expanded understanding of reality. The possibility of encountering life forms with different ethical systems could further challenge our own ethical frameworks, forcing us to reconsider our values and priorities.

The nature of contact could significantly impact the future trajectory of humanity. A peaceful and collaborative encounter could lead to a golden age of scientific progress and interspecies cooperation, bringing together diverse cultures and species for the betterment of all. However, a hostile encounter, characterized by conflict and competition for resources, could lead to a catastrophic outcome, jeopardizing the survival of humanity. The potential for miscommunication or misunderstanding is high, given the vast differences in communication styles, values, and technologies. Careful consideration and planning are crucial in developing effective strategies for interspecies communication and conflict resolution. International cooperation and diplomacy would be vital in navigating this complex and potentially dangerous situation.

Furthermore, the long-term implications extend beyond immediate contact. The discovery of extraterrestrial life could lead to the colonization of other planets and the expansion of human civilization into the cosmos. This would necessitate the development of advanced technologies for interstellar travel and the establishment of sustainable habitats on other planets. It would require us to confront the ethical challenges associated with planetary colonization, including the potential impact on indigenous extraterrestrial life and the preservation of biodiversity in new environments. These challenges will require careful consideration and a long-term perspective, prioritizing the well-being of both human and extraterrestrial life.

The economic impact could be transformative. The discovery and exploitation of extraterrestrial resources could lead to unprecedented economic growth, potentially eliminating poverty and scarcity. However, this could also lead to new forms of economic inequality and conflict, especially if access to these resources is unequally distributed. Careful planning and international cooperation will be necessary to ensure equitable access to these resources and to prevent the exacerbation of existing economic inequalities.

The political landscape could be fundamentally reshaped. The formation of international alliances and treaties for cooperation in space exploration and the management of extraterrestrial resources would be crucial. However, this could also lead to new geopolitical tensions and conflicts, especially if power imbalances or competing interests arise. International collaboration and diplomacy will be essential in navigating these complex challenges and maintaining peace.

Ultimately, the future of humanity in the face of extraterrestrial contact is uncertain. It holds both immense opportunities for progress and significant risks of conflict and catastrophe. The key to navigating this uncertain future lies in a careful and thoughtful approach, prioritizing international cooperation, ethical considerations, and long-term planning. Our ability to adapt and respond effectively will determine whether the discovery of extraterrestrial life leads to a golden age of human civilization or a period of profound upheaval and decline. The choices we make today will determine the trajectory of humanity's future for generations to come. The exploration of the cosmos is not merely a scientific endeavor, but a journey of self-discovery, forcing humanity to confront its own limitations, biases, and ultimately, its place in the grand scheme of the universe. This journey will test our resilience, our adaptability, and our collective capacity for empathy and understanding – qualities that will be essential in navigating the vast unknowns that lie ahead. The future, in its vastness, offers both exhilarating possibilities and daunting challenges, a future that remains unwritten, a story yet to be told, and dependent entirely on the decisions we make now.

| 11 |

Search for Biosignatures in Our Solar System

The search for life beyond Earth naturally leads us to Mars, our closest planetary neighbor. Its geological history, hinting at a warmer, wetter past, makes it a prime target in the quest for past biosignatures – remnants of life, however ancient. The compelling evidence for past liquid water on Mars, revealed through orbital imagery showcasing dried-up riverbeds, lake basins, and even potential ocean shorelines, fuels the ongoing exploration. These features, coupled with the discovery of subsurface ice, suggest that conditions conducive to life may have existed on Mars billions of years ago, potentially harboring microbial life.

The exploration of Mars has spanned decades, progressing from initial flybys and orbiters to sophisticated rovers capable of conducting detailed in-situ analysis. Early missions, like Mariner 4 in 1965, provided initial, albeit limited, images of a seemingly barren and cratered landscape. However, subsequent missions, particularly the Viking program of the 1970s, marked a significant turning point. Viking 1 and Viking 2, equipped with sophisticated instruments designed to detect microbial life, landed on the Martian surface and conducted experiments to analyze soil samples. While the results were inconclusive, sparking ongoing debate, they laid the foundation for future missions. The Viking experiments, though ultimately ambiguous regarding the existence of life, provided invaluable experience and data for the planning of subsequent endeavors. The ambiguity highlighted the complexities of detecting life on other planets, underscoring the need for more sophisticated and diverse approaches.

The years following Viking saw a period of relative quiet in Mars exploration, though data from orbiters continued to contribute to our understanding of the planet's geology and climate. The Mars Global Surveyor, launched in 1996, provided high-resolution images of the Martian surface, further strengthening the evidence for past water activity. This mission, alongside other orbital surveys, played a crucial role in selecting landing sites for future rovers, focusing on areas with the highest potential for preserving evidence of past life. The identification of such regions involved sophisticated image analysis techniques and careful consideration of geological processes, a testament to the increasingly advanced tools and methods used in planetary exploration.

The new millennium witnessed a resurgence of ambitious Mars exploration, with the launch of rovers capable of extensive surface exploration and sophisticated sample analysis. The Mars Exploration Rovers, Spirit and Opportunity, landed in 2004 and far exceeded their planned mission lifetimes. These rovers provided compelling evidence for a past environment conducive to life, detecting minerals formed in the presence of water. Opportunity, in particular, made headlines with its discovery of hematite, a mineral often associated with water-rich environments on Earth, significantly bolstering the case for a wetter Mars in the past. The rovers' longevity and resilience underscored the ingenuity of their design and the dedication of the mission teams involved.

The Curiosity rover, launched in 2011, marked another significant leap forward. Larger and more technologically advanced than its predecessors, Curiosity's advanced instrumentation allows for a more comprehensive analysis of Martian rocks and soil. Curiosity's key findings include the discovery of organic molecules – the building blocks of life – within Martian rocks. While not direct evidence of life, the presence of these molecules strongly suggests that the conditions necessary for life could have existed. Further analysis revealed that the Martian environment, at least in Gale Crater, Curiosity's landing site, may have been habitable for microbial life for extended periods. The sustained operation of Curiosity has expanded our knowledge of Mars significantly and has provided valuable data for planning future missions.

The Perseverance rover, launched in 2020, represents the most ambitious and sophisticated Mars rover mission to date. Its primary objective is not merely to search for signs of past life but also to collect and cache samples for potential return to Earth in the future. This mission marks a crucial step towards potentially making direct observations of Martian biosignatures under the intense scrutiny of Earth-based laboratories. The Ingenuity helicopter, accompanying Perseverance, has demonstrated the feasibility of aerial exploration on Mars, opening up exciting new possibilities for future missions. The Perseverance mission's targeted exploration of Jezero Crater, a former lakebed, maximizes the chances of finding preserved evidence of past life. The samples collected will be meticulously analyzed for a range of biosignatures, with the ultimate goal of definitively answering the question of whether life ever existed on Mars.

The Mars Sample Return (MSR) campaign, a joint effort between NASA and ESA, aims to bring these samples back to Earth. This complex and ambitious undertaking will involve multiple missions, spanning decades. The return of Martian samples represents an unprecedented opportunity to analyze them with sophisticated instruments far beyond the capabilities of any rover currently on Mars. This meticulous analysis will provide definitive answers regarding the presence of past life on Mars, potentially revolutionizing our understanding of the prevalence of life in the universe. The logistical complexities of the MSR mission, however, are immense, requiring precise coordination and advanced technology to successfully retrieve the samples and return them to Earth, minimizing contamination from terrestrial sources.

Beyond the rovers and landers, orbital missions continue to play a vital role in Mars exploration. Missions like the Mars Reconnaissance Orbiter (MRO) and the Trace Gas Orbiter (TGO) are providing detailed information about the Martian atmosphere and surface, contributing to our understanding of the planet's climate history and identifying potential targets for future surface missions. The high-resolution images provided by these orbiters allow scientists to select landing sites based on geological features of interest, ensuring that rovers focus their efforts on the most promising areas for finding evidence of past life. The data from these orbiters, integrated with the data from surface rovers, provides a comprehensive perspective on the Martian environment, facilitating more informed conclusions about its habitability.

The search for past life on Mars is a complex and multifaceted endeavor, involving scientists from diverse fields, from geology and geochemistry to microbiology and astrobiology. The integration of data from various missions, along with advanced analytical techniques, is crucial for interpreting the findings and drawing meaningful conclusions. The ambiguity of some past results underscores the challenges inherent in detecting ancient microbial life, necessitating continuous refinement of methodologies and instrumentation. The collaborative efforts of international space agencies are fundamental to advancing our knowledge about Mars and pushing the boundaries of our exploration.

The ongoing exploration of Mars and the search for past life represent a pivotal moment in humanity's pursuit of understanding our place in the universe. The potential discovery of past life on Mars would not only confirm that life has existed beyond Earth but would also provide invaluable insights into the origins and evolution of life itself. It would profoundly impact our understanding of the prevalence of life in the universe and profoundly influence future exploration strategies. The ethical implications of such a discovery would also require careful consideration, prompting discussions regarding the protection of potential Martian biosignatures and the responsible exploration of other celestial bodies. The journey to uncover the secrets of Mars and its past continues, driving innovation and pushing the boundaries of human ingenuity as we strive to answer one of science's most fundamental questions: Are we alone?

Europa, one of Jupiter's four Galilean moons, has emerged as a compelling target in the search for extraterrestrial life. Unlike Mars, where we primarily seek evidence of *past* life, Europa presents the intriguing possibility of a *present-day* ocean, hidden beneath a thick layer of ice. This subsurface ocean, shielded from the harsh radiation environment of Jupiter, could potentially harbor life.

The evidence for Europa's subsurface ocean is multifaceted and compelling. Initially, observations by the Voyager and Galileo spacecraft revealed a surprisingly smooth surface, lacking the extensive cratering expected on a body of Europa's age. This smoothness suggests a process of resurfacing, likely driven by tectonic activity associated with a liquid ocean beneath. Further analysis of Galileo's data, including measurements of Europa's magnetic field, revealed a strong interaction with Jupiter's magnetosphere. This induced magnetic field is consistent with a conductive subsurface layer, most plausibly a salty ocean of liquid water.

The thickness of Europa's icy shell remains a subject of ongoing investigation and debate. Estimates range from tens of kilometers to over 100 kilometers, with the uncertainty stemming from the limitations of remote sensing techniques. However, the consensus is that the ice shell is not monolithic; it is likely fractured and possibly contains internal pockets of liquid water. Furthermore, the possibility of cryovolcanism, where water erupts from the subsurface ocean onto the surface, cannot be ruled out. Such eruptions, if confirmed, would offer a direct pathway to sample the composition of the ocean without the need for drilling through the kilometers of ice. The potential for cryovolcanism, and associated plume activity, has led to detailed observations aiming to detect water vapor or other volatiles venting from the surface.

The composition of Europa's subsurface ocean is crucial to assessing its potential for life. Spectroscopic analyses of Europa's surface have detected salts, including sulfates and possibly chlorides, providing further evidence of a salty ocean. The presence of these salts is significant, as they could help maintain the ocean's liquid state at Europa's low temperatures, lowering its freezing point. Furthermore, the salts themselves could provide essential nutrients for any potential life forms within the ocean. The presence of other dissolved minerals, organic molecules, and potentially even hydrothermal vents on the ocean floor would further enhance the ocean's habitability, mimicking the environments that fostered the earliest life on Earth.

The potential energy sources for life in Europa's subsurface ocean are also being investigated. Tidal forces exerted by Jupiter's gravity cause significant flexing and heating within Europa's interior, providing a potential source of energy for hydrothermal vents. These vents, analogous to those found on the ocean floor on Earth, release heat and chemicals that could support chemosynthetic life – life that doesn't rely on sunlight for energy, but rather on chemical reactions. The existence of hydrothermal vents on Europa is a compelling possibility, offering a rich and diverse environment capable of supporting a complex ecosystem even in the absence of sunlight. Modeling studies suggest that the heat generated by tidal forces could create enough thermal energy to sustain a potentially vast network of hydrothermal vents, far surpassing the�� and density of those found on Earth.

Another potential source of energy for life on Europa comes from the interaction between the ocean and the overlying ice shell. The movement and melting of ice could lead to the release of chemicals and energy, creating a dynamic and habitable environment within the ice itself. The interplay between the ice and the ocean, influenced by tidal forces and other geological processes, creates a unique and potentially complex environment for the evolution of life. Furthermore, the radiation environment, while harsh at the surface, significantly diminishes with depth, creating a habitable zone beneath the icy shield where life could thrive.

The search for evidence of life on Europa necessitates advanced exploration strategies. The vast distance to Jupiter and the challenges posed by Europa's icy surface require innovative approaches. Future missions will likely involve orbiters equipped with advanced radar and magnetometers to further characterize the subsurface ocean and ice shell. These orbiters could also perform detailed spectroscopic analyses to search for signs of organic molecules or other biosignatures on Europa's surface. Ideally, future missions will also include landers capable of drilling through the ice shell to directly sample the ocean water and sediment beneath. This ambitious goal requires significant technological advancements and careful planning to minimize the risks of contamination.

The possibility of sending robotic submarines to explore Europa's ocean represents a significant long-term objective. Such a mission would allow for in-situ exploration of the ocean floor, potentially identifying hydrothermal vents, collecting samples, and searching for signs of life. However, the technological challenges associated with developing a submarine capable of surviving the extreme pressures and temperatures of Europa's ocean are considerable. Furthermore, the long journey to Europa and the limitations on communication with a remote submarine pose significant hurdles.

The implications of discovering life on Europa would be profound. It would not only confirm that life exists beyond Earth, but it would also dramatically alter our understanding of the prevalence of life in the universe and the conditions under which it can arise and thrive. The discovery of life on Europa would likely be a multi-cellular life form, unlike the likely past microbial life forms expected on Mars. The discovery would suggest that subsurface oceans beneath icy shells could be common throughout the universe, greatly expanding the range of environments considered potentially habitable. It would also encourage further exploration of icy moons in other star systems, potentially leading to the discovery of even more diverse life forms beyond our own solar system. The ethical implications of such a discovery would need careful consideration, highlighting the importance of responsible exploration and the protection of potential extraterrestrial life.

In summary, Europa represents a compelling target in the ongoing search for extraterrestrial life. The compelling evidence for a subsurface ocean, its potential for habitability, and the ongoing development of advanced exploration technologies all contribute to the excitement surrounding this enigmatic moon. While the challenges are significant, the potential rewards – a groundbreaking discovery that could redefine our understanding of life and its place in the cosmos – make Europa a priority destination for future exploration. The meticulous planning and technological innovation required for future missions underscore the importance of this endeavor and the scientific breakthroughs it promises to yield. The exploration of Europa is a testament to humanity's insatiable curiosity and our unwavering pursuit of knowledge about our place in the universe.

Enceladus, a small, icy moon of Saturn, has captivated the scientific community with its stunning displays of cryovolcanism. Unlike Europa, where the subsurface ocean remains hidden beneath a thick ice shell, Enceladus offers a tantalizing glimpse into its internal ocean through spectacular geysers erupting from fissures near its south polar region. These geysers, plumes of icy particles and water vapor, spray material hundreds of kilometers into space, providing a unique opportunity to analyze the composition of Enceladus's subsurface ocean without the need for the incredibly complex task of drilling through kilometers of ice.

The discovery of these geysers was a monumental moment in astrobiology. Data gathered by the Cassini spacecraft, which orbited Saturn from 2004 to 2017, revealed the existence of these powerful plumes. Cassini's instruments, including its Cosmic Dust Analyzer (CDA) and Ion and Neutral Mass Spectrometer (INMS), were able to directly sample the material within these plumes, providing unprecedented insights into their composition. The analysis revealed a surprising abundance of water vapor, alongside significant amounts of organic molecules, salts, and silicates. The presence of these components strongly suggests interaction with a rocky seabed, pointing to the presence of hydrothermal vents.

The detection of organic molecules, the building blocks of life, is particularly exciting. Cassini's INMS detected a wide array of organic compounds, including methane, carbon dioxide, and more complex hydrocarbons. While the exact composition and origin of these organics remain subjects of ongoing investigation, their presence undeniably enhances the potential habitability of Enceladus's subsurface ocean. The possibility that these molecules are directly related to hydrothermal vent activity further strengthens the case for a potentially biogenic source. The detection of complex organic molecules, unlike the simpler ones seen on some other celestial bodies, suggests a potentially more complex and dynamic chemical environment. Future missions with more sensitive instruments could help us uncover the specifics of these organic molecules and further probe the possibility of a biological origin.

The presence of salts and silicates in the plumes provides further evidence for hydrothermal activity. The salts, likely originating from the ocean floor, are consistent with the interaction of water with the rocky core. The silicates, tiny mineral particles, suggest that water is interacting with and altering the composition of the rocky seabed, providing further evidence of hydrothermal activity. The size and nature of the silicate particles themselves hold clues about the processes involved, hinting at the temperature and pressure conditions prevailing at the ocean floor and suggesting a powerful heat source.

Further bolstering the hypothesis of hydrothermal vents is the observed energy budget of Enceladus. The moon's internal heat, seemingly insufficient to account for the observed level of geological activity from tidal forces alone, suggests additional energy sources might be at play. Hydrothermal vents, fuelled by chemical reactions between the water and the rocky interior, are an ideal candidate for this additional energy source. These vents could provide both the chemical energy necessary for chemosynthetic life and the warm, stable environment required for its sustenance. The amount of heat emitted by these vents, even considering their potential size, suggests a considerable overall impact on the energy budget. Advanced models are currently being used to study the details of Enceladus's internal heat flow and better understand the interactions between tidal forces, radioactive decay, and the chemical processes occurring within the hydrothermal vents.

The detection of molecular hydrogen (H2) in the plumes provides compelling evidence of hydrothermal activity. Molecular hydrogen is a byproduct of hydrothermal vent activity on Earth, where water interacts with rocks at high temperatures, undergoing chemical reactions that release hydrogen. The detection of molecular hydrogen in the Enceladus plumes, via Cassini's INMS data, is strong evidence supporting the hypothesis of hydrothermal vents capable of providing a significant energy source. This H2 is especially exciting because it can support a vast network of microorganisms in a chemosynthetic ecosystem, analogous to hydrothermal vent communities found in the deep ocean on Earth. The levels of H2 detected suggest that it could indeed fuel a substantial biosphere, providing the necessary energy and nutrients for life to thrive.

The implications of these discoveries are profound. The presence of a subsurface ocean, hydrothermal vents, organic molecules, and molecular hydrogen combine to create an environment that is remarkably similar to those that fostered the earliest life forms on Earth. While there is no direct evidence of life on Enceladus yet, the potential for its existence is compellingly high. The fact that we have detected these ingredients of life without even needing to penetrate the icy crust speaks to the tantalizing possibility that life might be present.

Future missions to Enceladus will be crucial in further investigating the potential for life. A dedicated mission, equipped with more sophisticated instruments, is needed to further analyze the composition of the plumes, to search for biosignatures directly within the plume material, and even to potentially sample the subsurface ocean itself. Advanced sensors and instruments would allow for more detailed characterization of the organic molecules, revealing their structure and complexity. They could also detect more subtle biosignatures, which might be present in trace amounts.

One particularly ambitious goal for future missions involves sending a lander to directly sample the plumes or, even more ambitiously, a probe to penetrate the ice shell and explore the ocean directly. A lander could collect samples of the ice particles and analyze them in situ, searching for signs of life or further evidence of hydrothermal activity. This would allow for a far more comprehensive analysis than is possible with remote sensing alone. However, the challenges of designing a lander or probe that can withstand the harsh environment of Enceladus and communicate its findings back to Earth are considerable.

A subsurface probe would be an immensely challenging undertaking. The journey to Saturn, the challenges of landing on a small, icy moon, and the need for a system capable of autonomously drilling through the ice shell and navigating the ocean floor require a significant technological leap. Moreover, extreme precautions must be taken to avoid contaminating Enceladus's pristine ocean with terrestrial microbes.

The discovery of extraterrestrial life, particularly in an environment as seemingly inhospitable as Enceladus's subsurface ocean, would be revolutionary. It would confirm that life can thrive in a wide variety of environments, dramatically altering our understanding of the prevalence of life in the universe and the conditions necessary for its existence. It would also demonstrate the incredible resilience and adaptability of life, challenging our assumptions about what constitutes a habitable world. The implications of such a discovery would extend far beyond the scientific community, reshaping our perspective on our place in the cosmos and our responsibility to protect potentially fragile ecosystems beyond our planet. The exploration of Enceladus, driven by scientific curiosity and technological innovation, is a testament to humanity's ongoing quest to understand our place within the vastness of the universe.

Moving beyond the icy realms of Enceladus, our search for extraterrestrial biosignatures leads us to another captivating moon in the Saturnian system: Titan. Unlike Enceladus's subsurface ocean of water, Titan boasts a strikingly different environment, one dominated by liquid hydrocarbons. This moon presents a unique and compelling case for the possibility of life based on alternative chemistries, challenging our Earth-centric biases in the search for life beyond our planet.

Titan's atmosphere, significantly denser than Earth's, is composed primarily of nitrogen, with methane playing a crucial role in shaping its surface and climate. The methane cycles through various phases, existing as a gas in the atmosphere, forming clouds, and precipitating as liquid methane rain, carving rivers and filling lakes on the surface. These lakes and seas, primarily composed of liquid methane and ethane, form a vast network across Titan's polar regions, creating a landscape unlike anything else in our solar system. The abundance of hydrocarbons makes Titan's surface remarkably different from Earth's, providing a unique laboratory for studying the potential for life to thrive in a hydrocarbon-rich environment.

The exploration of Titan has been significantly advanced by the Cassini-Huygens mission. The Huygens probe, which landed on Titan's surface in 2005, provided the first close-up images and data from the moon's surface, revealing a surprisingly complex and dynamic landscape. Images showed evidence of erosion, suggesting that liquid methane plays a significant role in shaping the surface. Huygens also measured the composition of the atmosphere and the surface, providing crucial information for understanding Titan's unique environment. Subsequent analysis of data gathered by Cassini, orbiting Saturn throughout the early 2000s, has further expanded our knowledge. Data from radar instruments revealed the extent of Titan's hydrocarbon seas and lakes, mapping their depths, morphology, and even the presence of potential islands. This information is crucial for understanding the movement of these hydrocarbon fluids and their potential role in shaping Titan's geology.

The existence of liquid methane on Titan's surface raises the intriguing possibility of a methane-based biochemistry. While life as we know it relies on water as a solvent, the unique properties of liquid methane could potentially support different forms of life. Methane, though less efficient as a solvent than water, can dissolve certain organic molecules, and could theoretically serve as a medium for chemical reactions that might be the basis of a metabolic process. The low temperatures on Titan, however, present significant challenges. Chemical reactions generally proceed much slower at low temperatures, which could limit the rate of metabolic processes and thus the potential for complex life.

However, it is important to note that our understanding of what constitutes "life" is still largely based on Earth's life forms. The possibility of life arising from entirely different chemical processes, particularly in a low-temperature, hydrocarbon-rich environment like that on Titan, must be seriously considered. The presence of diverse organic molecules in Titan's atmosphere, as detected by Cassini, is a critical factor in these considerations. Though primarily simpler compounds like methane and ethane, there is evidence suggesting the presence of more complex organic molecules, possibly formed through atmospheric photochemical reactions. These complex molecules, if present in sufficient quantities and with appropriate properties, could be the basis for a unique biochemistry not yet understood. Future missions to Titan will need the capability of resolving the fine details of these atmospheric molecules to fully assess their complexity.

Another critical element to consider is the potential energy sources available on Titan. On Earth, the primary energy source for life is sunlight, but on Titan, sunlight is significantly weaker due to the dense atmosphere. However, other potential energy sources could support life, including:

Chemical energy: Reactions between different hydrocarbons could provide energy for metabolic processes, particularly in the subsurface. The interaction between liquid methane and other compounds present in the crust could release energy suitable for life. The relatively recent discovery of potential cryovolcanism on Titan suggests the possibility of subsurface geological activity that could provide both heat and chemical energy sources.

Tidal energy: As with many moons in the solar system, Titan experiences tidal forces from Saturn, which could generate internal heat and potentially contribute to hydrothermal activity, similar to the processes thought to be occurring on Enceladus. While this would be a weaker energy source than on more volcanically active bodies, it might nonetheless provide sufficient energy for certain biological processes.

Radioactive decay: Similar to Enceladus, radioactive decay within Titan's core could contribute to internal heating, providing an additional energy source for subsurface processes. The details of Titan's internal structure and heat flow are still subjects of active investigation, but radioactive decay is a likely significant contributor.

The search for biosignatures on Titan presents unique challenges. Our traditional methods for detecting life are largely based on identifying molecules associated with water-based life. On Titan, the focus shifts to detecting molecules associated with methane-based metabolisms, if such metabolisms exist. This requires development of novel methods and instruments specifically designed to detect and identify potentially novel biological markers. This includes the development of more sensitive analytical techniques and theoretical models to predict the potential properties of these biomarkers.

Future missions to Titan will be pivotal in advancing our understanding of this unique moon. These missions should involve more sophisticated instruments capable of studying both Titan's atmosphere and surface in greater detail, and also more sophisticated subsurface analysis instruments. A combination of orbital observations and, ideally, landers and probes capable of sampling Titan's surface and subsurface environments will be vital in this endeavor. Advanced robotic missions could gather samples of various materials, providing a far more comprehensive analysis than is possible with remote sensing alone. This would involve the collection of samples from the surface materials, including lake sediments, and potentially even accessing the subsurface using specialized drills or probes designed to navigate the complex terrain.

Crucially, these missions must be designed with utmost care to avoid contaminating Titan's pristine environment with terrestrial microbes. Strict planetary protection protocols are essential to ensure that any future discoveries are genuinely extraterrestrial in origin and not artifacts of Earth contamination.

The exploration of Titan represents a significant step beyond our traditional search for extraterrestrial life. It forces us to reconsider our understanding of what constitutes a "habitable" environment and what forms life might take beyond Earth. The possibility of life based on alternative chemistries, thriving in a hydrocarbon-rich environment, could redefine our understanding of the prevalence of life in the universe, demonstrating the remarkable resilience and adaptability of life in unexpected forms. The potential for discovering life on Titan is undoubtedly a compelling driver for continued exploration and innovation in astrobiology, propelling our quest for answers to one of humanity's most profound questions – are we alone?

The search for life beyond Earth necessitates a multifaceted approach, encompassing both the refinement of existing technologies and the development of entirely novel strategies for exploration. While missions like Cassini-Huygens have provided invaluable insights into the environments of Titan and Enceladus, future endeavors demand a significant leap forward in our capabilities. This involves not only more advanced instrumentation, but also a fundamental shift in our thinking about what constitutes a habitable environment and how we might detect life outside the familiar context of Earth.

One of the most crucial aspects of future missions will be the enhanced ability to conduct in-situ analysis. While remote sensing techniques, such as spectroscopy, have proven invaluable in identifying potential biosignatures from orbit, they are inherently limited in their resolution and the types of information they can provide. To overcome this limitation, future missions will need to incorporate advanced robotic systems capable of collecting samples and performing detailed analyses on the surface or even within the subsurface of these celestial bodies. This involves the development of miniaturized, robust laboratory systems capable of performing a variety of analyses, including chemical characterization, isotopic analysis, and the search for organic molecules with potential biological significance.

For example, missions to Enceladus could involve a lander equipped with a drill to access the subsurface ocean, possibly deploying a submersible probe to explore the ocean's depths. This probe could collect samples of water and ice, analyze their chemical composition, and search for signs of microbial life or other biosignatures. Such a mission would require significant advancements in robotics, power sources, and communication technologies to operate effectively in the harsh environment of Enceladus, particularly its extreme cold and the high pressures encountered at great depths. Moreover, designing a probe capable of enduring this intense environment, while simultaneously being sterile enough to prevent Earth contamination, presents a significant engineering challenge.

Similarly, future missions to Titan could involve more sophisticated landers and rovers capable of traversing the diverse terrain, including the vast hydrocarbon lakes and seas. These rovers would be equipped with a range of instruments, including advanced spectrometers, mass spectrometers, and microscopes, to analyze the composition of the surface materials, the atmosphere, and even the potential subsurface regions. They could analyze the complex organic molecules that exist in Titan's atmosphere and lakes, searching for patterns suggestive of biological activity. These rovers would need to be capable of autonomously navigating the challenging terrain of Titan, withstanding the extreme temperatures and pressure. The potential for cryovolcanism on Titan also introduces a significant complexity to mission planning, requiring careful consideration of the potential hazards associated with geothermal activity.

Beyond landers and rovers, the development of aerial platforms such as drones or blimps could provide a revolutionary approach to exploring Titan's diverse landscapes. These platforms could offer a unique perspective, providing high-resolution imagery and enabling sampling of various locations across the surface with relative ease. This strategy would particularly benefit from improvements in autonomous navigation and flight control, especially considering the dense atmosphere and potential wind conditions on Titan. These advancements could enable more efficient exploration of vast areas and enhance the scope of sampling capabilities.

In addition to surface exploration, future missions will also benefit significantly from improved orbital observations. This could include the deployment of advanced telescopes and spectrometers capable of detecting subtle changes in the atmospheres of these celestial bodies. These instruments would be designed to search for volatile organic compounds indicative of biological activity, focusing not just on detection, but on the identification of complex organic patterns, potentially indicating the presence of living organisms. Improved orbital sensors might also reveal clues about the subsurface activity of these moons, such as plumes erupting from Enceladus or subtle shifts in Titan's surface topography indicative of ongoing geological processes. Analyzing these subtle indicators could further constrain models of habitability and provide valuable context for the planning of future landed missions.

Moreover, advancements in computing power and artificial intelligence (AI) will be essential for analyzing the vast amounts of data generated by these future missions. AI-powered algorithms could be used to identify potential biosignatures in complex datasets, reducing the workload for scientists and allowing for the exploration of large parameter spaces for biological signals that may otherwise go unnoticed. The capabilities of AI in autonomously analyzing and classifying this complex data will play a critical role in prioritizing samples for more detailed analysis by the human researchers on Earth.

The development of new technologies specifically designed for the search for extraterrestrial life is crucial. This includes the development of novel sensors, instruments, and analysis techniques capable of detecting and identifying potentially unfamiliar biosignatures. It's important to bear in mind that life beyond Earth may be vastly different from life on Earth, and our understanding of biosignatures must be expanded to incorporate such possibilities. This necessitates a multidisciplinary approach, drawing on expertise from various fields such as chemistry, biology, physics, and engineering to devise innovative solutions.

The ethical implications of searching for extraterrestrial life must also be carefully considered. The discovery of life beyond Earth would have profound implications for humanity, and it is essential to develop responsible and ethical guidelines for the exploration and study of such life. This includes not only the protection of extraterrestrial environments from contamination by Earth-based life, but also the ethical considerations of engaging with any potentially discovered life forms. The concept of planetary protection demands paramount consideration, implementing stringent protocols to prevent contamination of potentially habitable environments by Earth-based microbes, thereby ensuring that any life detected is genuinely extraterrestrial in origin. Furthermore, any interaction or engagement with an extraterrestrial life form would necessitate a thoughtful and well-considered approach, governed by ethical guidelines that prioritize the preservation of both the discovered life and the human pursuit of knowledge.

Finally, international collaboration is crucial for the success of future missions. The exploration of other celestial bodies, especially those that may harbor life, requires the pooling of resources and expertise from around the world. This necessitates a coordinated effort between space agencies, research institutions, and individual scientists. Such collaborations enhance the capabilities of individual nations by combining expertise, reducing redundancy and optimizing the effective use of limited resources. The challenges of exploring distant worlds are significant, requiring vast investment and long-term commitment. International collaboration not only offers economic benefits, but also promotes intellectual exchange and a shared sense of purpose in exploring the profound question of whether life exists elsewhere in the universe. These combined efforts ensure a robust and sustainable approach to the exploration of space, promising more efficient and effective missions while fostering a spirit of international unity.

| 12 |

The Drake Equation and Probability of Life

The quest to answer the fundamental question of whether we are alone in the universe has driven scientific inquiry for decades. While direct evidence of extraterrestrial life remains elusive, the development of sophisticated tools and theoretical frameworks has allowed us to approach this question with a new level of rigor. One such framework, profoundly influential in the field of astrobiology, is the Drake Equation. Developed by Frank Drake in 1961, this equation is not a definitive calculation but rather a thought experiment, a probabilistic argument designed to stimulate discussion and guide research into the prevalence of intelligent life in the cosmos.

The Drake Equation is deceptively simple in its appearance, yet its implications are profound and far-reaching. It attempts to estimate *N*, the number of active, communicative extraterrestrial civilizations in our galaxy that we might potentially detect. The equation takes the form:

$$N = R_{*} \times f_p \times n_e \times f_l \times f_i \times f_c \times L$$

Let's dissect each term individually:

R⊕ (Star formation rate): This represents the average rate of star formation in our galaxy. Astronomical observations have provided relatively precise estimates for this value, indicating a substantial number of stars are born in the Milky Way each year. The rate fluctuates over time, and the specific value used can vary depending on the model and data utilized in the estimation. However, a generally accepted range is between one and several stars formed per year. The uncertainty in this value highlights the inherent limitations of relying on currently available data and the constant refinement that occurs within the astronomical community as more powerful telescopes and analytical techniques are developed.

f_p (Fraction of stars with planetary systems): This term represents the proportion of stars that possess planetary systems. Recent exoplanet discoveries, particularly those made by the Kepler and TESS missions, have revolutionized our understanding of planetary systems. These discoveries reveal that planets are remarkably common, orbiting a vast number of stars across a wide range of sizes and orbits. The initial estimates, based on less comprehensive data, significantly underestimated this fraction. Current data suggests that a substantial portion, possibly the majority, of stars likely have planets. While this figure is becoming increasingly precise, uncertainties still exist regarding the detection limits of current technologies and the potential for undiscovered planet types.

n_e (Number of planets per solar system suitable for life): This term addresses the number of planets within each stellar system that possess conditions potentially conducive to the emergence of life. This is a particularly complex factor, as the definition of "suitable for life" itself is constantly evolving. We are becoming increasingly aware of the diversity of environments that might support life, ranging from subsurface oceans on icy moons to planets with radically different atmospheric compositions from Earth. The value of n_e is subject to considerable debate and uncertainty, with estimates ranging from a few planets to only one within a given solar system. Our understanding of habitability is intrinsically bound to our knowledge of the diverse forms of life on Earth, highlighting the necessity to continue broadening this understanding to extrapolate more accurately into different environments.

f_l (Fraction of suitable planets where life actually appears): This represents the probability that life will actually emerge on a planet deemed suitable for its development. This is perhaps the most challenging term to estimate, primarily because we only have one data point – Earth. While we are beginning to unravel the complexities of abiogenesis (the origin of life), we lack a comprehensive understanding of the precise conditions and processes that led to life's emergence on our planet. It is crucial to remember that the emergence of life might be a rare occurrence, or alternatively, might be a relatively common event given suitable conditions. The uncertainty in this term drives much of the debate surrounding the Drake Equation, necessitating the exploration of diverse potential pathways for the emergence of life.

f_i (Fraction of life-bearing planets where intelligent life emerges): This term considers the probability that life, once established, will evolve into intelligent forms. Evolution is a complex, non-linear process, subject to a multitude of chance events and environmental pressures. The emergence of intelligence, as we understand it, may be a relatively rare outcome, or it may be a more common occurrence than currently thought. Estimating this value necessitates the consideration of diverse evolutionary pathways and the inherent stochasticity of biological processes. We are only beginning to develop a deeper understanding of evolutionary pathways on Earth, limiting our ability to predict with any precision the likelihood of comparable developments in different extraterrestrial contexts.

f_c (Fraction of civilizations that develop communication technology): This represents the proportion of intelligent civilizations that develop the technological capacity to communicate across interstellar distances. This is a particularly speculative aspect of the equation, dependent on technological advancements and societal choices within each civilization. Some civilizations might choose not to communicate, or their technologies might not be suitable for interstellar communication. The development of sustainable advanced technology might be incredibly rare or relatively easy to achieve; current observations do not provide enough information to offer a definitive answer.

L (Lifetime of a communicating civilization): This is the average length of time that a civilization actively communicates across interstellar distances. This factor incorporates various factors, including technological lifespan, societal stability, and susceptibility to self-destruction (e.g., through nuclear war or environmental collapse). This is another highly speculative term, and estimates vary drastically. The longevity of technological civilizations might be short, or perhaps some civilizations might endure for thousands or millions of years. Our limited understanding of the factors driving the stability of advanced civilizations makes this term one of the largest sources of uncertainty in the equation.

The Drake Equation, in its totality, reveals more about the limitations of our current knowledge than about the precise number of extraterrestrial civilizations. However, its enduring value lies in its ability to focus research on the individual factors within the equation, encouraging scientific progress in fields such as astronomy, astrobiology, and planetary science. The ongoing search for exoplanets, the study of abiogenesis, and the exploration of the long-term sustainability of civilizations – all are directly influenced by the framework provided by the Drake Equation. It remains a powerful tool to organize our thinking and direct our efforts in answering one of humanity's most profound and enduring questions. While the numerical result remains highly speculative, its influence on the direction of scientific inquiry is undeniable.

The Drake Equation, while deceptively simple in its algebraic form, encapsulates a breathtakingly complex tapestry of probabilities and unknowns. Each term within the equation represents a significant hurdle in our quest to understand the prevalence of extraterrestrial life, and accurately estimating each parameter remains a challenge that continues to drive scientific exploration. Let's delve deeper into the individual components, acknowledging the considerable uncertainties inherent in each.

R◈ (Star Formation Rate): While astronomers have made significant strides in measuring the star formation rate in our galaxy, this value is not static. The rate varies over time, influenced by factors such as the availability of interstellar gas and dust, the gravitational interactions within the galactic structure, and the presence of active galactic nuclei. Current estimates range from one to several stars formed per year, but these figures are based on observations of currently visible star formation and extrapolations based on models of galactic evolution. These models themselves carry uncertainties, reflecting the complexity of the processes involved and the limitations of our observational tools. More distant star formation is harder to detect accurately, potentially leading to underestimation. The improvement of telescope technology, particularly in infrared astronomy, allowing us to peer through interstellar dust clouds, is crucial to refining this parameter. Further advancements in computational modeling techniques, allowing more detailed simulations of galactic dynamics, will further refine our understanding of this fundamental aspect of galactic evolution and the rate of star formation.

f_p (Fraction of Stars with Planetary Systems): The discovery of thousands of exoplanets has dramatically shifted our understanding of planetary systems, moving from a somewhat parochial view of our solar system as typical to an appreciation of the incredible diversity and prevalence of planets orbiting other stars. Missions like Kepler and TESS have revolutionized the field, revealing that planets are significantly more common than previously imagined. However, the very act of detection biases our understanding. We are more likely to detect larger planets closer to their stars due to the limitations of current detection techniques. Smaller, further planets remain much harder to discover, creating uncertainty in estimating the true fraction of stars with planetary systems. Technological advancements, such as the development of more sensitive instruments and novel detection techniques, are essential to overcome these limitations. The use of advanced statistical methods, accounting for observational biases and potential detection limits, will be crucial in refining our estimate of f_p, bringing us closer to a more accurate representation of the prevalence of planetary systems across the Milky Way.

n_e (Number of Planets per Solar System Suitable for Life): This parameter highlights the central challenge of defining habitability. Our understanding of what constitutes a habitable planet is intimately tied to our knowledge of life on Earth. We tend to favor planets similar to Earth in size, composition, and orbital distance from their star – but this might be a narrow-minded perspective. The discovery of extremophiles on Earth, organisms thriving in seemingly inhospitable environments like deep-sea hydrothermal vents or highly acidic lakes, expands the possibilities of where life could exist. Subsurface oceans on icy moons like Europa and Enceladus further broaden the definition of a habitable environment. The search for biosignatures – chemical indicators of past or present life – in the atmospheres of exoplanets could provide critical data, but interpreting these biosignatures will require a deep understanding of planetary evolution and atmospheric chemistry across a wide variety of conditions. The value of n_e, therefore, hinges on ongoing research into the limits of life and the potential for life to arise and persist in environments vastly different from Earth's.

f_l (Fraction of Suitable Planets Where Life Actually Appears): This represents the probability of abiogenesis

– the transition from non-living matter to living organisms. This is a fundamental question that remains stubbornly elusive, despite decades of research in fields like origin-of-life chemistry and astrobiology. Our single data point – Earth – is insufficient to make a robust statistical inference. While numerous hypotheses exist about the conditions conducive to abiogenesis, none are definitive. Did life emerge rapidly, or was it a rare and improbable event? The search for prebiotic molecules in meteorites and the study of extremophiles on Earth provide valuable clues. However, a definitive understanding of abiogenesis likely requires a combination of advanced laboratory experiments designed to simulate prebiotic conditions, improved modeling of early Earth's environment, and potentially the discovery of extant extraterrestrial life to provide more data points for statistical analysis.

f_i (Fraction of Life-Bearing Planets Where Intelligent Life Emerges): The evolution of intelligence is a complex process, not necessarily a predetermined outcome of the emergence of life. We only know of one instance – the evolution of intelligent life on Earth – and drawing broad conclusions from a single data point is fraught with peril. Moreover, our definition of "intelligence" is anthropocentric, limiting our ability to recognize other potential forms of intelligence that might emerge through different evolutionary trajectories. Could other forms of intelligence exist that are vastly different from our own? Could technology develop through a completely different path? The evolutionary pathways leading to intelligence might be exceedingly rare, or perhaps intelligence is a common outcome given sufficient time and the right conditions. This remains a significant area of uncertainty, necessitating further research into evolutionary biology, particularly the study of complex adaptation and the factors that influence cognitive development.

f_c (Fraction of Civilizations That Develop Communication Technology): This term acknowledges the challenges of technological development and the potential for societal factors to influence a civilization's propensity for communication. A civilization might develop advanced technology without ever having the desire or capacity for interstellar communication. Technological choices are influenced by factors such as resource availability, societal structure, and even potential catastrophic events that could hinder development. The development and maintenance of technological capabilities requires enormous energy resources and stability, both of which might be limited factors for many civilizations. Understanding the factors influencing a civilization's trajectory, especially on long timescales, will require a more profound understanding of the complex interplay between technological advancement, societal structures, and environmental influences.

L (Lifetime of a Communicating Civilization): This parameter is perhaps the most speculative, reflecting the inherent uncertainty surrounding the long-term sustainability of technologically advanced civilizations. Several potential factors could lead to the demise of a civilization, including internal conflict, resource depletion, environmental catastrophe, or even self-annihilation through advanced weaponry. Our own civilization's relatively short history of advanced technology offers little in the way of long-term predictability. The study of historical trends in civilizations, the modeling of societal collapse, and the assessment of the long-term effects of environmental change could provide insights, though the long timescale involved makes this a formidable challenge. Understanding the factors that contribute to the longevity of a civilization remains crucial for a comprehensive understanding of the factors underlying the Drake Equation.

In conclusion, while the Drake Equation provides a powerful framework for thinking about the prevalence of extraterrestrial life, the sheer magnitude of uncertainty surrounding each term highlights the limitations of our current understanding. However, the equation's lasting value lies not in providing a precise numerical answer, but in its capacity to stimulate research and guide our efforts in the quest to answer one of humanity's most profound and enduring questions. The continuous advancement of observational astronomy, the deepening of our understanding of abiogenesis and evolutionary biology, and the exploration of factors influencing societal stability and technological development – all are directly fueled by the compelling questions raised by the Drake Equation.

Estimating the number of civilizations using the Drake Equation is akin to navigating a star chart riddled with nebulae and black holes. The equation, as we've seen, is a product of seven parameters, each shrouded in significant uncertainty. This translates into a breathtakingly wide range of potential answers, from a lonely Earth, the sole beacon of life in a vast, silent cosmos, to a galaxy teeming with intelligent civilizations.

Let's examine some of the attempts to plug values into the Drake Equation and the resulting estimates. It's crucial to remember that these estimations are highly speculative, reflecting the current state of our knowledge and the inherent uncertainties in each parameter. Different researchers will employ different values based on their own interpretations of the available data and their perspectives on the likelihood of various events.

One of the earliest and perhaps most optimistic estimations was made by Frank Drake himself in 1961. His initial calculation, using relatively generous values for each parameter, yielded a substantial number of communicating civilizations in our galaxy. This early estimate ignited a wave of enthusiasm within the scientific community, suggesting that we might not be alone. However, this early optimism was tempered by the uncertainties inherent in many of the parameters.

Subsequent estimates have varied dramatically, depending on the values chosen for each term. Some estimations have been remarkably conservative, suggesting only a handful, or even just one, communicating civilization in our galaxy. This more pessimistic outlook often reflects a greater emphasis on the challenges of abiogenesis, the rarity of intelligence, and the fragility of advanced civilizations. These assessments might highlight the potential for catastrophic events to curtail a civilization's lifespan or the difficulty in maintaining long-term technological advancement and interstellar communication capabilities.

The variations in estimated values of the parameters themselves are significant drivers of this range in final results. For instance, the star formation rate (R) varies based on differing interpretations of galactic evolution models and the completeness of star surveys. The fraction of stars with planetary systems (fp) has evolved significantly with the detection of thousands of exoplanets, yet the detection biases remain, affecting our confidence in the completeness of our catalogs of planetary systems.

The number of planets per solar system suitable for life (ne) is arguably the most uncertain term. As our understanding of extremophiles on Earth continues to expand, the very definition of "habitable" becomes increasingly ambiguous. Do we limit our search to Earth-like planets, or should we consider planets orbiting different types of stars, planets with subsurface oceans, or those with drastically different atmospheric compositions? A broadening of the definition of habitability dramatically impacts the value assigned to ne.

The probability of life appearing on a suitable planet (fl) is another enormous hurdle. We have only one data point: Earth. This inherent limitation makes any assessment of the frequency of abiogenesis highly speculative. The current lack of demonstrable extraterrestrial life, despite extensive searches, tends to push the estimates for fl towards the lower end of the range, reflecting a possible rarity of the processes that lead to life's origin. Furthermore, the timeframe for abiogenesis remains unclear. Was it a relatively rapid event, occurring shortly after the formation of a suitable planet, or did it take billions of years to occur? The answer to this question profoundly impacts the estimation of fl.

The probability of the emergence of intelligent life (fi) is equally uncertain. We again rely on the single case of Earth, making any estimation incredibly speculative. The trajectory of evolution is highly contingent on a multitude of factors, many of which remain poorly understood. Did the development of intelligence in humans follow a unique evolutionary pathway, or is it a more common outcome under certain conditions? The answer to this question will significantly impact our estimation of fi and, subsequently, the total number of intelligent civilizations.

The fraction of intelligent civilizations that develop communication technologies (fc) introduces further uncertainties. A civilization might develop advanced technologies without ever seeking to communicate with others. Technological choices might be influenced by social structures, resource allocation, and potential risks of interstellar communication. The development and maintenance of advanced communication technologies necessitate considerable energy and resource commitment. Civilizations might simply lack the capacity or the motivation to establish interstellar communications, diminishing the value of fc.

Finally, the lifetime of a communicating civilization (L) is perhaps the most challenging parameter to estimate. Several factors could lead to the demise of a civilization, including self-annihilation through conflict or environmental degradation, resource depletion, or catastrophic events. The comparatively short period of human technological advancement provides minimal data to make reliable predictions about the longevity of advanced civilizations. Furthermore, we do not know how long a civilization remains detectable through interstellar communication. A civilization might exist for millennia but only transmit signals for a relatively short time, reducing our chances of detecting them.

Given the vast uncertainties in these parameters, it's not surprising that estimations of the number of civilizations vary from a single civilization (ourselves) to millions or billions within the Milky Way galaxy alone. These estimates highlight the need for further research in many scientific areas, from the study of exoplanets and abiogenesis to the modeling of societal dynamics and technological development.

The true power of the Drake Equation lies not in providing a definitive answer, but in illuminating the immense gaps in our knowledge and in motivating further scientific inquiry. Each parameter poses fascinating scientific questions that drive research in diverse fields, underscoring the inherent interdisciplinary nature of the search for extraterrestrial intelligence. The exploration of these questions, through both theoretical modeling and observational astronomy, will gradually refine our estimations and bring us closer to answering one of humanity's most fundamental questions: Are we alone? The wide range of possibilities remains a powerful testament to the vastness of the cosmos and the profound mysteries that await discovery. Even with the inherent uncertainties, the very act of grappling with these questions pushes the boundaries of human knowledge and forces a re-evaluation of our place in the universe.

The Drake Equation, despite its inherent uncertainties, serves as a powerful tool, not just for estimating the number of extraterrestrial civilizations, but more importantly, as a framework for guiding research and focusing our efforts in the field of astrobiology. Its true significance lies not in generating a precise numerical answer – a feat currently impossible given our limited knowledge – but in its ability to highlight the crucial unknowns and stimulate research across numerous scientific disciplines. Consider it a scientific roadmap, charting a course through the uncharted territories of cosmic life.

The equation's seven parameters – star formation rate (R), fraction of stars with planetary systems (fp), number of planets per solar system suitable for life (ne), probability of life appearing on a suitable planet (fl), probability of intelligent life emerging (fi), fraction of intelligent civilizations developing communication technologies (fc), and the lifetime of a communicating civilization (L) – represent key areas of inquiry demanding further investigation. Each term compels us to delve deeper into the fundamental processes governing the origin and evolution of life, the formation and characteristics of planetary systems, and the dynamics of societal and technological development.

For instance, the parameter R, *representing the rate of star formation in our galaxy, is constantly being refined as our understanding of stellar evolution improves. Advanced telescopes, such as the Gaia mission, are providing increasingly accurate data on stellar populations, their ages, and their distribution. This refined data, in turn, leads to a more nuanced estimation of R,* directly impacting the overall result of the Drake Equation. Similarly, advancements in exoplanet detection techniques, including transit photometry and radial velocity measurements, are continually updating our estimates for fp, the fraction of stars with planetary systems. The Kepler mission and the Transiting Exoplanet Survey Satellite (TESS) have already revolutionized our understanding of planetary systems, revealing a stunning diversity of exoplanets, some orbiting within the habitable zones of their stars. However, observational biases still exist; we are more likely to detect larger planets closer to their stars, potentially skewing our estimation of fp.

The parameter ne, the number of planets per solar system suitable for life, poses perhaps the most significant challenge. The definition of "habitable" itself is continually evolving. While early estimations primarily focused on Earth-like planets orbiting Sun-like stars within the "Goldilocks zone" – the region where liquid water could exist on the surface – recent discoveries have expanded our horizons. The existence of extremophiles on Earth – organisms thriving in extreme environments like hydrothermal vents, acidic hot springs, and subglacial lakes – suggests that life might flourish under a far broader range of conditions than previously imagined. Planets with subsurface oceans, tidally locked planets, and those orbiting different types of stars, including red dwarfs, now fall under the umbrella of potentially habitable worlds. This expanded definition of habitability drastically increases the potential value of ne, significantly impacting the overall result of the Drake Equation.

The parameters fl and fi, representing the probability of life arising and the probability of intelligent life evolving, are perhaps the most speculative. We have only one data point – Earth – to inform our estimations. This lack of comparative data leads to a wide range of possibilities, from life's emergence being a relatively common event to an extraordinarily rare occurrence. Further research in abiogenesis – the study of the origin of life – is crucial to refining our estimations. Experiments exploring prebiotic chemistry, simulating early Earth conditions, and analyzing the potential for life to emerge in other environments are essential to understanding the probability of life originating elsewhere. Similarly, research on evolutionary biology, particularly on the factors driving the emergence of intelligence, is essential to better understand the parameter fi. Are there specific evolutionary pathways that inevitably lead to intelligence, or is the development of intelligence a highly contingent and rare event?

The parameter fc, the fraction of intelligent civilizations that develop communication technologies, introduces another layer of complexity. A civilization might possess advanced technology without ever choosing to communicate with other civilizations. Factors such as societal structures, resource allocation, and potential risks associated with interstellar communication could influence this decision. Furthermore, even if a civilization chooses to communicate, the technologies used might be beyond our current capacity to detect.

Finally, the parameter L, the lifetime of a communicating civilization, is profoundly uncertain. The relative brevity of human civilization's technological advancement offers limited insight into the potential longevity of advanced civilizations. Factors such as environmental degradation, resource depletion, internal conflict, or catastrophic events could limit the lifespan of a civilization. Furthermore, the duration for which a civilization actively transmits detectable signals might be considerably shorter than its overall lifetime.

In conclusion, the Drake Equation's significance lies not in providing a precise answer to the question of extraterrestrial life, but in serving as a dynamic framework for focused research. It highlights the critical unknowns in astrobiology, motivating research across various disciplines and prompting a deeper understanding of the universe's potential for life. The broad range of potential answers, from a lonely Earth to a galaxy teeming with life, remains a testament to the vastness of the cosmos and the profound mysteries that still await discovery. The pursuit of answers, driven by the Drake Equation's inherent uncertainties, fuels scientific inquiry and pushes the boundaries of human knowledge, ultimately enriching our understanding of our place in the universe. The ongoing research, spurred by this compelling equation, promises a future where the once-unanswerable question of our cosmic solitude may finally find its response. Each incremental advancement in our understanding of the equation's parameters brings us closer to a more complete and accurate assessment, even if a definitive answer remains elusive. The journey itself, a testament to human curiosity and scientific endeavor, is perhaps as significant as the destination.

Refining our understanding of the Drake Equation requires a multi-pronged approach, focusing on both theoretical advancements and technological improvements. The inherent uncertainties embedded within each parameter necessitate a concerted effort across diverse scientific disciplines. While we may never arrive at a precise numerical answer, the process of refinement itself yields invaluable insights into the fundamental processes governing the origin, evolution, and persistence of life in the universe.

One crucial area requiring further investigation is the parameter *R* – the rate of star formation in our galaxy. While current estimates provide a reasonable approximation, the accuracy of these estimations relies heavily on the completeness of our stellar catalogs and our understanding of stellar evolution. Ongoing surveys like Gaia are revolutionizing our knowledge, providing more precise measurements of stellar distances, luminosities, and ages. However, challenges remain, particularly concerning the detection of low-mass stars and brown dwarfs, which may contribute significantly to the overall star formation rate but remain difficult to detect with current technology. Future observational efforts focusing on these faint objects, combined with advanced computational modeling of stellar populations, are essential for refining *R*. Moreover, a deeper understanding of star formation processes in different galactic environments – spiral arms, galactic centers, dwarf galaxies – is needed to account for the variations in star formation rates across the galaxy. This will lead to more nuanced estimations of *R*, thereby enhancing the accuracy of the Drake Equation.

The parameter f_p, the fraction of stars with planetary systems, is another area where substantial progress has been made, yet much remains unknown. The Kepler mission and TESS have dramatically increased the number of known exoplanets, revealing a remarkable diversity in planetary systems – from hot Jupiters to super-Earths, and everything in between. However, detection biases continue to influence our understanding. Current methods are more sensitive to larger planets located closer to their stars, potentially leading to an underestimation of smaller, potentially habitable planets further from their host stars.

Future missions with improved sensitivity and broader observational coverage are crucial to overcome these biases. Furthermore, a more comprehensive understanding of planetary formation mechanisms is essential. Numerical simulations and theoretical models can provide valuable insights into the processes that govern planet formation and the distribution of planets within various types of stellar systems. Combining these theoretical advancements with observational data from next-generation telescopes will greatly refine our understanding of f_p.

The parameter n_e, the number of planets per solar system suitable for life, presents perhaps the greatest challenge. Our understanding of habitability is constantly evolving. The discovery of extremophiles on Earth – organisms thriving in extreme environments – has expanded our definition of habitable conditions, suggesting that life may exist beyond the traditional "Goldilocks zone." Planets with subsurface oceans, tidally locked planets, and those orbiting red dwarf stars are now considered potentially habitable. Furthermore, the chemical composition of a planet's atmosphere, the presence of a magnetic field, and the planet's geological activity all play crucial roles in determining its habitability. Advanced atmospheric characterization techniques, using spectroscopy to analyze the composition of exoplanet atmospheres, will be crucial in identifying biosignatures and determining the potential habitability of extrasolar planets. Moreover, sophisticated climate models are needed to simulate the long-term evolution of planetary climates under different conditions. This interdisciplinary approach, integrating observations, laboratory experiments, and theoretical modeling, will lead to a more comprehensive and accurate assessment of n_e.

The parameters f_l and f_i, representing the probability of life arising and the probability of intelligent life evolving, remain highly speculative. We have only one data point – Earth – to inform our estimates, making these parameters some of the most challenging to constrain. Research in abiogenesis – the study of the origin of life – is crucial to address f_l. Laboratory experiments simulating prebiotic chemistry, exploring the potential for life to originate in different environments, and studying the chemical and physical conditions necessary for life's emergence are essential steps. Analyzing the potential for life to originate in other environments, such as subsurface oceans or other planetary bodies within our solar system, could also provide valuable insights. For f_i, research in evolutionary biology and artificial intelligence can contribute significantly. Understanding the evolutionary pressures that led to the emergence of intelligence on Earth and exploring the potential for alternative evolutionary pathways could help constrain this parameter. However, inherent uncertainties remain, emphasizing the limitations of extrapolating from a single example.

The parameter f_c, the fraction of intelligent civilizations that develop communication technologies, introduces further complexity. A civilization might possess advanced technology without ever choosing to broadcast its existence to the universe. Sociological and anthropological studies might offer insights into the factors that influence a civilization's decision to communicate, including resource allocation, societal structures, and potential risks associated with interstellar contact. Moreover, even if a civilization does attempt to communicate, the technologies used might be beyond our current detection capabilities. This necessitates the development of increasingly sensitive and sophisticated detection methods, ranging from radio telescopes to the search for technosignatures – signs of advanced technology – in the cosmos.

Finally, *L*, the lifetime of a communicating civilization, is perhaps the most uncertain parameter. The relatively short history of human civilization offers limited insight into the potential longevity of advanced civilizations. Factors such as environmental degradation, resource depletion, internal conflicts, or catastrophic events could significantly limit a civilization's lifespan. Historical studies of past civilizations, combined with studies of long-term sustainability, can provide some insights. However, this parameter remains inherently speculative, underscoring the vast uncertainties inherent in the Drake Equation.

In conclusion, while the Drake Equation might not yet offer a precise answer to the question of extraterrestrial life, its value lies in its ability to guide research and inspire collaborative efforts across scientific disciplines. The process of refining its parameters continues to drive technological advancements, enhance our understanding of fundamental processes, and expand our knowledge of the universe's potential for life. Each incremental improvement in our understanding, regardless of whether it leads to a definitive numerical answer, remains a significant contribution to the ongoing quest to unravel one of humanity's most profound questions. The journey, fueled by scientific curiosity and driven by the uncertainties inherent in the equation, is itself a testament to human ingenuity and our relentless pursuit of knowledge.

13

Interstellar Travel and Communication

The sheer scale of interstellar distances presents the most immediate and daunting obstacle to interstellar travel. Even the closest star system, Alpha Centauri, lies approximately 4.37 light-years away – a distance so vast that it defies human comprehension. To put this in perspective, consider that a light-year is the distance light travels in one year, approximately 9.46 trillion kilometers. Traveling to Alpha Centauri at the speed of our fastest spacecraft today would take tens of thousands of years, rendering such a journey impractical, if not impossible, within the lifespan of a human generation. This vastness necessitates the development of propulsion systems capable of achieving speeds significantly exceeding anything currently within our technological reach.

One of the major impediments to achieving such speeds is the energy requirement. According to Einstein's famous equation, $E=mc^2$, energy and mass are equivalent. To accelerate a spacecraft to a significant fraction of the speed of light requires an immense amount of energy, far beyond anything we can currently generate and harness. Consider the energy required to accelerate even a relatively small spacecraft to just 10% the speed of light. The kinetic energy would be astronomical, demanding energy sources far exceeding our current capabilities. This challenge extends beyond the initial acceleration phase; maintaining velocity and decelerating upon arrival at the destination require equally significant energy expenditures.

Several theoretical propulsion systems have been proposed to address this challenge. Nuclear fusion, for instance, offers a potentially powerful energy source, but mastering controlled fusion remains a significant hurdle. While significant progress has been made in fusion research, achieving sustained fusion reactions with a net energy output sufficient for interstellar travel is still far from reality. Other concepts, such as antimatter propulsion, are even more speculative. Antimatter, when annihilated with matter, releases a tremendous amount of energy, theoretically exceeding the energy output of nuclear fusion. However, the production and containment of antimatter are enormous technological challenges, requiring significant advancements in both physics and engineering.

Beyond the energy challenges, the technological hurdles involved in interstellar travel are numerous and profound. Building a spacecraft capable of withstanding the rigors of interstellar travel presents a major engineering feat. The spacecraft would need to withstand extreme temperature fluctuations, high levels of radiation, and the impacts of interstellar dust and particles. Shielding the crew from these hazards necessitates the development of advanced materials and sophisticated protective systems. Furthermore, the spacecraft's life support systems would need to function flawlessly for decades, if not centuries, ensuring the survival of the crew throughout the journey. This involves self-sufficiency in terms of air, water, food, waste recycling, and medical care. Designing a closed-loop ecosystem capable of sustaining human life over such extended periods is an immense technological challenge, requiring deep understanding of ecological dynamics and advanced bioengineering.

Another critical aspect is the navigation and control of the spacecraft during an interstellar voyage. Accurate navigation over interstellar distances requires extremely precise measurements and sophisticated guidance systems. Maintaining communication with Earth over such vast distances poses another significant challenge. The time delay inherent in interstellar communication, due to the finite speed of light, would mean that any communication between the spacecraft and Earth would be delayed by years, or even decades, depending on the distance. This lag in communication complicates the mission control and decision-making processes.

Moreover, the psychological impact of long-duration interstellar travel on the crew needs careful consideration.

The isolation, confinement, and extended periods away from family and loved ones could lead to psychological distress and interpersonal conflicts. Addressing these potential challenges necessitates the development of effective countermeasures, such as crew selection based on psychological resilience, the implementation of advanced communication technologies to maintain contact with Earth, and the provision of opportunities for social interaction and psychological support within the spacecraft.

The challenges extend beyond the technical and psychological realms to encompass ethical and societal considerations. The immense costs associated with interstellar travel would require significant global cooperation and resource allocation, potentially diverting resources from other pressing societal needs. Furthermore, the implications of encountering extraterrestrial life are far-reaching and potentially profound, raising ethical questions regarding our interactions with other civilizations and the potential consequences of interstellar contact. The ethical frameworks governing such interactions need careful deliberation and international consensus.

The feasibility of interstellar travel hinges on the successful integration of numerous scientific and technological advancements across multiple disciplines. Progress in materials science, energy production, propulsion systems, life support technology, navigation and communication systems, and even our understanding of human psychology is critical. While the challenges are formidable, the pursuit of interstellar travel remains a compelling endeavor, driving scientific innovation and pushing the boundaries of human ingenuity. The successful accomplishment of interstellar travel would mark a significant leap in human civilization, transforming our understanding of the universe and our place within it. However, achieving this milestone requires sustained effort, international collaboration, and a clear-eyed assessment of the immense challenges that lie ahead. The journey will undoubtedly be fraught with obstacles, but the potential rewards are immeasurable. The exploration of interstellar space and the potential discovery of extraterrestrial life represent a profound step in humanity's evolution.

Building upon the immense technological and logistical hurdles previously discussed, let's delve into the proposed methods for interstellar travel. These solutions range from incremental advancements to radical departures from our current understanding of physics, each with its own set of challenges and potential breakthroughs.

One of the more realistic, albeit still extremely challenging, approaches is **fusion propulsion**. Nuclear fusion, the process that powers the sun, involves combining light atomic nuclei to form heavier ones, releasing vast amounts of energy in the process. Unlike nuclear fission, which splits atoms, fusion produces significantly less radioactive waste. The prospect of harnessing this energy for propulsion is alluring, as it promises a far more powerful and efficient energy source than chemical rockets. However, achieving controlled and sustained fusion reactions on a scale suitable for interstellar travel remains a monumental task. Current fusion reactors, while demonstrating the feasibility of the process, are far from producing a net energy output that could be effectively used for propulsion. The engineering challenges associated with containing and manipulating the extremely high temperatures and pressures required for fusion remain considerable. Moreover, the development of lightweight, efficient fusion reactors capable of withstanding the rigors of interstellar travel presents significant material science challenges. Research into advanced materials, such as high-temperature superconductors, is crucial for developing efficient and durable fusion reactors for spacecraft. Despite these significant obstacles, fusion propulsion remains a leading candidate for interstellar travel due to its theoretically higher energy density compared to chemical rockets.

Another theoretical propulsion system that holds immense promise, though remains largely speculative, is **antimatter propulsion**. Antimatter is the counterpart of ordinary matter, possessing the same mass but opposite charge. When antimatter and matter collide, they annihilate each other, releasing enormous amounts of energy – far exceeding the energy output of even nuclear fusion. Theoretically, a small amount of antimatter could propel a spacecraft to a significant fraction of the speed of light. However, the production and containment of antimatter pose significant challenges. Producing antimatter requires immense amounts of energy, and its storage presents a major obstacle, as it must be kept isolated from ordinary matter to prevent annihilation. The development of effective containment systems capable of holding antimatter for extended periods is a key hurdle. Furthermore, the efficient conversion of the annihilation energy into thrust also needs to be addressed. While the energy density is unparalleled, the technological hurdles in production, storage, and controlled release currently render antimatter propulsion a long-term prospect, relegated to the realms of advanced theoretical physics. The sheer cost and technological complexity make it a distant dream, at least for the foreseeable future.

Moving into even more speculative territory, we have **warp drives**, a concept that has captured the imaginations of science fiction enthusiasts for decades. Based on theoretical solutions to Einstein's field equations, warp drives involve warping spacetime itself, creating a bubble around the spacecraft that allows for faster-than-light travel without violating the principles of relativity. This concept involves the manipulation of exotic matter, a hypothetical form of matter with negative mass-energy density, which is currently unconfirmed and possibly nonexistent. The energy requirements for a warp drive are predicted to be astronomical, possibly exceeding the total energy output of the sun. Even if exotic matter were to exist and could be harnessed, the technological challenges in manipulating spacetime itself are beyond our current understanding. While warp drives represent a fascinating area of theoretical physics, their feasibility remains highly questionable, bordering on the improbable with our current knowledge.

Beyond these primary propulsion methods, other less developed concepts warrant consideration. **Laser propulsion**, for instance, involves using powerful lasers located on Earth or in space to propel a spacecraft. This eliminates the need to carry large amounts of fuel onboard, potentially reducing the spacecraft's mass and increasing its velocity. However, the development of lasers powerful enough to accelerate a spacecraft to relativistic speeds remains a considerable challenge. The energy requirements are enormous, and the precision required to maintain the beam over interstellar distances is extremely demanding. Further advancements in laser technology and efficient energy transfer mechanisms are necessary to explore this option fully.

Nuclear pulse propulsion, another theoretical approach, involves detonating nuclear bombs behind a spacecraft, utilizing the resulting shockwave for propulsion. This concept, while powerful, has obvious drawbacks, raising significant safety and environmental concerns. The potential for radioactive contamination and the political challenges of weaponizing space make it a highly controversial approach.

A less technologically ambitious, but potentially more practical, approach lies in the realm of **generation ships**. These are enormous spacecraft designed to house multiple generations of humans during an interstellar voyage. The journey would take centuries or millennia, with generations of humans born, living, and dying on the ship. This approach bypasses the need for extraordinarily fast speeds, but it introduces a host of unique challenges related to maintaining the ship's internal ecosystem and ensuring the psychological well-being of generations raised in a confined environment. Such a concept requires a deep understanding of ecological dynamics, advanced life support systems, and sophisticated methods for managing the societal and psychological needs of a large population confined to a closed environment for potentially hundreds or thousands of years.

Finally, it's worth noting that advancements in other areas of science and technology are crucial for interstellar travel, irrespective of the propulsion system chosen. Advancements in materials science are necessary to create lighter, stronger, and more radiation-resistant spacecraft. Improved life support systems capable of providing sustenance for the crew for extended periods are also essential. Furthermore, sophisticated navigation and communication systems are vital for guiding the spacecraft and maintaining contact with Earth.

The pursuit of interstellar travel is a monumental undertaking, demanding breakthroughs in multiple scientific and engineering disciplines. While many of the proposed methods remain firmly in the realm of theory, ongoing research and technological advancements continue to slowly chip away at these immense challenges. The journey to the stars is long and arduous, but the potential rewards—a deeper understanding of the universe and the potential discovery of extraterrestrial life—make it a pursuit worthy of continued dedication and global cooperation. The road is paved with significant obstacles, but the destination holds a profound allure that continues to drive human ingenuity and inspire ambitious endeavors into the cosmos. The exploration of the interstellar medium, therefore, requires a sustained and multi-faceted approach, encompassing not only propulsion methods but also the development of robust life support, reliable communication, and effective strategies for managing the significant challenges of long-duration spaceflight.

Even if we overcome the monumental hurdles of interstellar travel, the challenge of communication across the vast expanse of interstellar space remains a formidable obstacle. The sheer distances involved dwarf even our most ambitious propulsion systems. Even at a significant fraction of the speed of light, journeys to even the nearest stars would take decades, centuries, or even millennia. This temporal gap presents a significant communication challenge, introducing considerable delays in transmitting and receiving messages. A simple "hello" sent to a nearby star system could take years, or even decades, to reach its destination, and the response would take an equally long time to return. This latency significantly impacts the feasibility of real-time communication, demanding alternative approaches to information exchange.

The vast distances also affect the strength of signals. Electromagnetic radiation, the most practical method for interstellar communication, weakens as it travels through space. The inverse square law dictates that signal strength diminishes proportionally to the square of the distance. Therefore, a signal sent to a distant star system will arrive significantly attenuated, requiring incredibly powerful transmitters and sensitive receivers. The interstellar medium itself is not a perfect vacuum; it contains low-density gas and dust that can further scatter and absorb electromagnetic radiation, increasing the challenges of signal transmission and reception. To overcome this attenuation, we would need extraordinarily powerful transmitters, capable of broadcasting messages at exceptionally high energies, and exquisitely sensitive receivers capable of detecting the faintest signals amid the cosmic noise. The development of such advanced technologies is a significant engineering endeavor, demanding breakthroughs in areas such as antenna design, signal processing, and radio astronomy.

Furthermore, the challenge extends beyond the mere transmission of a signal; the decoding of an alien message poses an equally daunting problem. Assuming we receive a signal, deciphering its content requires overcoming the fundamental barrier of differing communication systems. Humans utilize complex languages structured around grammar, syntax, and semantics, which are culturally defined and may be uniquely human. There is no guarantee that an extraterrestrial civilization would communicate using a system remotely similar to our own. Their language could be entirely different, based on visual signals, sound frequencies beyond our auditory range, or even entirely novel methods of information transfer that we cannot even imagine. Even if the signal were in a format we recognize, such as radio waves or electromagnetic pulses, deciphering its meaning would require sophisticated computational linguistics and potentially years or decades of analysis. This underscores the need for developing robust methods of signal analysis and artificial intelligence algorithms capable of pattern recognition and language decipherment, particularly within a vast and chaotic data landscape of potential signals. The challenge extends beyond the technical decoding to the semantic understanding; we would need to not only identify patterns but also interpret their meaning within a wholly alien cultural context.

The nature of the message itself is also an unknown variable. An extraterrestrial civilization might transmit simple mathematical equations as a form of universal language, aiming to establish contact through a common foundation in mathematics and physics. Alternatively, they could send complex cultural messages, potentially encoding information about their biology, technology, or societal structures, posing a challenge to anthropologists, linguists, and even sociologists attempting to interpret the data. This interdisciplinary nature of interstellar communication highlights the need for a collaborative approach, involving scientists from diverse backgrounds working in concert to decipher the message's content.

The technological challenges related to interstellar communication are multifaceted and interconnected. Firstly, the power requirements for transmission are substantial. To reach interstellar distances, signals need to be exceptionally powerful, potentially demanding significant advances in energy generation and transmission technologies. This might involve the development of novel energy sources or highly efficient energy conversion mechanisms. Secondly, the bandwidth of the signal is crucial; a higher bandwidth allows for faster data transmission, facilitating a more efficient exchange of information. This requires advancements in signal processing and modulation techniques. Thirdly, the design and construction of incredibly large and sensitive antennas are essential for detecting faint interstellar signals. This demands progress in materials science, engineering, and potentially even novel approaches to antenna design that leverage physics beyond our current understanding.

Beyond technological hurdles, the fundamental question of whether we should attempt interstellar communication remains a topic of much debate. The potential risks, as highlighted by the "Wow!" signal and the Fermi Paradox, cannot be overlooked. Contact with a more advanced civilization could have unforeseen consequences, either beneficial or catastrophic. Some scientists and thinkers advocate for caution, suggesting a period of careful observation and analysis before attempting active transmission. Others argue that the potential scientific and cultural rewards outweigh the risks. This ethical consideration adds a layer of complexity to the practical challenges of interstellar communication.

The search for extraterrestrial intelligence (SETI) plays a crucial role in interstellar communication. SETI projects, using large radio telescopes and other detection methods, actively scan the skies for signs of alien signals. These projects have yielded both intriguing anomalies and frustrating silence, highlighting the difficulty of detection and the sheer immensity of the search space. Advancements in SETI techniques, including more sensitive receivers, wider frequency ranges, and more sophisticated signal processing algorithms, are necessary to improve our chances of detecting extraterrestrial communication. The development of advanced AI algorithms that can sift through vast amounts of data and identify potential signals is particularly critical.

Given the many challenges, a phased approach is essential. Initially, our focus should be on improving our signal detection capabilities through further investment in SETI and the development of more sensitive technologies. Parallel research into more effective message encoding and decoding techniques is vital. This will require intensive interdisciplinary collaboration between mathematicians, computer scientists, linguists, and astrobiologists. Simultaneously, we must consider the ethical implications of interstellar communication, establishing a framework for responsible contact and risk mitigation strategies.

In conclusion, interstellar communication presents a set of immense challenges. While the distances, signal attenuation, and linguistic complexities pose formidable obstacles, the potential rewards are too significant to ignore. The journey towards interstellar communication demands not only significant technological advancements but also a careful consideration of ethical considerations and strategic planning for a field rich in both promise and peril. Our future engagement with the cosmos depends on overcoming these hurdles, and the rewards of success would undoubtedly reshape humanity's understanding of its place in the universe. The path forward involves a multi-pronged approach, combining technological innovation with a clear-eyed assessment of potential risks and ethical responsibilities. Only then can we hope to truly engage in the interstellar dialogue that holds the potential to transform our world.

The challenges of interstellar communication, as previously discussed, are monumental. However, the quest to bridge the vast cosmic gulf necessitates exploring diverse communication methodologies. While the limitations of speed and signal strength remain significant, the potential for breakthroughs in several areas offers a glimmer of hope. Let's delve into some of the most promising methods currently under consideration.

The most established and widely explored method for interstellar communication is the use of radio waves. This approach leverages our existing understanding of radio astronomy and utilizes electromagnetic radiation within a frequency range that can penetrate the interstellar medium with relatively minimal attenuation. Radio waves offer a substantial advantage in terms of cost-effectiveness and technological maturity; many radio telescopes are already in operation, and the technology for transmitting and receiving radio signals is relatively well-developed. Moreover, the ability of radio waves to penetrate interstellar dust and gas makes them a suitable carrier for signals across vast interstellar distances. SETI projects, as mentioned earlier, largely rely on radio wave detection to search for extraterrestrial signals. However, radio waves also have their drawbacks. Their relatively low bandwidth limits the rate of data transfer, meaning it could take an extremely long time to send even moderately sized messages. The low energy density of radio waves requires large and powerful antennas to achieve the necessary signal strength for interstellar communication.

A more energy-efficient alternative to radio waves is the use of lasers. Laser communication systems offer a significantly higher bandwidth than radio waves, allowing for faster data transfer and potentially more complex message encoding. The tightly focused beam of a laser also minimizes signal dispersion, reducing the power needed to transmit a signal over long distances. However, laser communication is more challenging than radio communication due to its precise alignment requirements. Maintaining a stable and accurate laser beam pointing towards the target over interstellar distances necessitates sophisticated technological advancements in beam steering and tracking mechanisms. Moreover, the laser's narrow bandwidth, while offering advantages in data transfer rate, can be easily absorbed or scattered by interstellar dust and gas, making it less robust than radio waves in certain environments. Therefore, the choice between radio waves and lasers for interstellar communication will depend on factors such as the distance to the target, the available bandwidth, and the tolerance for signal attenuation.

Beyond radio waves and lasers, more speculative methods for interstellar communication are being explored. One such approach is the use of neutrinos. Neutrinos are subatomic particles that have extremely weak interactions with matter, allowing them to travel through vast distances without significant attenuation. Their ability to penetrate the interstellar medium essentially unhindered makes them a potentially ideal carrier for interstellar messages. However, the technology for generating and detecting neutrino signals is still in its infancy, and the incredibly low probability of neutrino interactions requires extremely powerful sources and incredibly sensitive detectors. The immense energy requirements to generate a detectable neutrino signal pose a significant technological barrier, making it currently impractical for interstellar communication.

Another futuristic communication method is the utilization of gravitational waves. These ripples in spacetime, predicted by Einstein's theory of general relativity, are exceptionally difficult to detect, but they possess the remarkable property of being unaffected by interstellar dust and gas. This makes them theoretically capable of traveling unimpeded across interstellar distances. However, the technological challenges associated with generating and detecting gravitational waves are insurmountable with current technology. Furthermore, generating a gravitational wave with sufficient energy to transmit information over such vast distances would require technology far beyond our present capabilities. Hence, gravitational wave communication remains a purely theoretical concept at this stage.

The choice of communication method is not a one-size-fits-all proposition. Several factors need to be considered when selecting the optimal approach. These include the distance to the target, the desired data transfer rate, the available power, the technological maturity of the method, and the tolerance for signal attenuation. A cost-benefit analysis must be performed to determine the most appropriate strategy based on the specific circumstances.

Furthermore, the communication methodology needs to consider the potential characteristics of extraterrestrial civilizations. Assuming we successfully establish contact, it's crucial to understand that an alien civilization's technological capabilities and communication preferences might be vastly different from our own. Their communication methods may utilize entirely different forms of energy or even exploit phenomena we haven't yet discovered. This necessitates developing robust and flexible communication systems capable of detecting and interpreting a wide range of signals.

The development of sophisticated artificial intelligence (AI) plays a critical role in interstellar communication. AI algorithms will be indispensable for analyzing vast amounts of data from SETI projects and identifying potential signals amidst the noise. Machine learning techniques will be crucial for recognizing patterns and deciphering the meaning of potentially alien messages. AI will also play a crucial role in developing advanced signal encoding and decoding techniques, potentially bridging the gap between vastly different communication systems. This interdisciplinary approach, combining advancements in signal processing, AI, and linguistics, is essential for successfully deciphering alien communication.

Ethical considerations also warrant careful attention. Before initiating active interstellar communication, we must thoroughly evaluate the potential risks and rewards. The "Wow!" signal, briefly mentioned before, serves as a reminder that even ambiguous signals can spark intense scientific interest and speculation. Conversely, encountering a more advanced civilization might lead to unintended and potentially harmful consequences. A thoughtful and cautious approach, possibly involving a period of passive observation before active transmission, is warranted.

The development of a standardized communication protocol is another vital aspect of interstellar communication. This protocol should be based on universal principles, such as mathematics and physics, to enhance the chances of understanding. Using simple mathematical principles as a common language could help overcome the limitations of differing cultural contexts and languages. However, even fundamental concepts like numbers could be encoded in entirely different ways by another civilization. The challenge of establishing a truly universal communication protocol requires a deep understanding of mathematics and the limits of mathematical thought, ensuring our approach is both universal and comprehensible.

Ultimately, the path to interstellar communication is a long and challenging one. It requires significant advancements in several scientific and technological domains, as well as careful consideration of ethical and philosophical issues. However, the potential scientific and cultural rewards of establishing contact with extraterrestrial intelligence are too significant to ignore. The journey ahead involves a combination of relentless innovation, interdisciplinary collaboration, and a mindful approach to the potential consequences, representing a collective endeavor for humanity's future in the vast cosmic ocean.

The sheer distances involved in interstellar travel and communication introduce a fundamental constraint: time. While we might readily discuss light-years as a unit of distance, truly grasping the implications of this scale for communication and travel requires a shift in perspective, a move away from our human-centric understanding of time. A message sent to even the nearest star system, Proxima Centauri, takes over four years to arrive, and the response another four years to return. This eight-year round trip is not merely a technological hurdle; it's a profound temporal chasm that reshapes our notions of real-time interaction and immediate feedback.

Consider the challenges of a simple exchange of information. A question posed today might receive an answer eight years hence. This delay isn't merely an inconvenience; it fundamentally alters the nature of conversation. Imagine trying to engage in a lively debate, a collaborative scientific project, or even a casual chat across such a temporal gulf. The conversational flow would be agonizingly slow, interrupted by years of silence punctuated by delayed responses, rendering any sense of spontaneity impossible. The context of the initial message might entirely change in the intervening years, making the delayed response irrelevant or even meaningless.

This time delay also impacts the very nature of scientific inquiry. Any collaborative research project across interstellar distances would require meticulous planning and exceptional patience. Experiments would need to be designed with the long delay in mind, potentially necessitating years of planning and data accumulation before any meaningful results could be exchanged. The slow pace of communication could hamper progress, limiting the speed at which scientific understanding is advanced. The time delay also affects the very act of observation. We observe distant celestial objects as they were in the past, not as they are in the present. Light from the Andromeda galaxy, for instance, takes millions of years to reach us, meaning we see it not as it is now, but as it was millions of years ago. This observational delay extends to any potential interstellar communication, complicating our understanding of the sender's present state.

Furthermore, the vast distances and the time delays significantly impact interstellar travel. Even with hypothetical advanced propulsion systems capable of near-light speed travel, the journey to even the closest star systems would take years, decades, or even centuries. The human lifespan, typically measured in decades, suddenly becomes a limiting factor in interstellar exploration. A mission to a distant star might require multiple generations of astronauts, potentially leading to sociological and psychological challenges on board the interstellar vessel. Generational ships, where individuals are born, live, and die during the course of the voyage, present a range of intricate logistical and social problems requiring considerable foresight. The very concept of a 'mission' takes on a vastly different meaning when stretched across such a long timescale.

The time scales involved in interstellar communication also raise profound questions about the nature of civilizations involved. If we were to communicate with a civilization located thousands of light-years away, we would be communicating with a civilization that existed thousands of years in the past. What guarantees remain that this civilization still exists? Has their society changed drastically during those millennia? Have they experienced technological breakthroughs or catastrophic collapses we can only speculate about? Indeed, they may not even exist in their previous form when our message finally arrives, rendering our communication futile or leading to misinterpretations.

To illustrate the profound implications of these time scales, consider the following hypothetical scenario. We detect a signal from a civilization located 1000 light-years away. We reply immediately, but it will take 1000 years for our response to reach them. If they were to reply back, it would then take another 1000 years for their message to reach us. The entire exchange would take a minimum of 2000 years, a timescale spanning multiple generations of our civilization. By the time we receive their response, much of the context of our original message might be lost to history, making comprehension of their response very challenging. Technological advancements, societal shifts, and linguistic evolution in the interim could render understanding their reply virtually impossible.

Furthermore, the time scale inherent in interstellar communication challenges our anthropocentric view of time. The 'now' we experience is not a universal 'now'. The concept of simultaneity, straightforward on a terrestrial scale, becomes highly relative when dealing with interstellar distances. There is no single, shared 'present' across vast cosmic stretches. This calls for a significant reframing of how we approach interstellar communication, emphasizing patience, long-term planning, and the acceptance of inherent temporal ambiguities. The very act of communication transforms from a real-time interaction into a process spanning millennia, requiring a profoundly different understanding of exchange and reciprocity.

The profound implications of time in interstellar communication extend beyond practical challenges. It calls into question our own place within the cosmic timeline, inviting a humbling reflection on the vastness of time and the ephemeral nature of our own existence. It invites introspection on the nature of our assumptions about other civilizations, reminding us that our understanding of extraterrestrial life might be vastly limited by our own temporal perspective. The vast time scales involved are not merely technological hurdles but fundamental realities that shape the very nature of interstellar interaction. They invite us to consider the patience, persistence, and perhaps even serendipity required to establish contact and truly bridge the cosmic chasm of time and distance. Ultimately, understanding these time scales is crucial not only for the practicalities of interstellar communication but also for framing our understanding of our own place within the universe.

| 14 |

The Societal Impact of Discovery

The sheer scale of a discovery of extraterrestrial life—a discovery that would irrevocably alter our understanding of our place in the universe—possesses the potential to transcend the usual limitations of international relations and foster unprecedented global cooperation. The unifying power of a shared, awe-inspiring experience, particularly one that shatters long-held assumptions about humanity's uniqueness, could serve as a powerful catalyst for international unity and collaboration. The immediate effect would likely be a dramatic shift in global priorities. Nationalistic rivalries, while undoubtedly persisting to some degree, would likely take a backseat to the urgent need for collective action in response to this extraordinary event. The initial stages would likely involve a global information-sharing effort, driven by a sense of shared wonder and the need to rapidly analyze and understand the implications of the discovery. International scientific collaborations, already prevalent in certain fields, would be exponentially intensified, pooling the world's brightest minds to decipher the mysteries surrounding extraterrestrial life.

Imagine the rapid mobilization of resources. The kind of technological and financial investment required to understand and potentially communicate with an extraterrestrial civilization would dwarf any single nation's capabilities. The necessity for coordinated global action would become self-evident, demanding the creation of robust international frameworks for research, communication, and resource allocation. National governments, even those with historically antagonistic relationships, would find themselves compelled to collaborate on a scale unseen before. The development of sophisticated communication technologies designed to bridge the interstellar divide would necessitate a massive, globally coordinated research effort. Expertise from diverse fields – from radio astronomy and astrophysics to linguistics, anthropology, and even philosophy – would be essential, highlighting the interconnectedness of knowledge and the need for diverse perspectives.

This global cooperation wouldn't be limited to the scientific realm. The economic implications of a discovery of extraterrestrial life are potentially immense. The discovery of new technologies, resources, or even fundamentally new scientific principles could reshape global economies, prompting the need for equitable distribution of benefits and the development of new international economic frameworks to manage this sudden influx of transformative knowledge. International treaties and agreements, possibly under the aegis of the United Nations, would be necessary to regulate the use of these new resources and prevent exploitation or conflict. Furthermore, ethical considerations would necessitate global discourse and consensus-building.

Ethical dilemmas surrounding potential interaction with an extraterrestrial civilization are numerous and complex. Questions about the potential risks, both to ourselves and to the extraterrestrial life form, would necessitate carefully considered global policies to prevent accidental harm. The potential for conflict, driven by differing cultural values or competing national interests, must be carefully addressed through open international dialogue and the development of universally accepted ethical guidelines. Furthermore, the protection of extraterrestrial life would become a major international concern, likely demanding the creation of global agencies dedicated to safeguarding the newly discovered life forms and their environment. This would necessitate an international consensus on principles of planetary protection, ensuring that humanity's actions don't irrevocably harm this extraterrestrial life.

Beyond immediate practical concerns, the discovery of extraterrestrial life promises a profound shift in humanity's self-perception. The long-held anthropocentric view of our place in the universe would be decisively challenged, forcing a reevaluation of our values and priorities. This shift in perspective could potentially foster a renewed sense of global unity, emphasizing our shared humanity in the face of the vast cosmos. The experience of confronting our place within a universe teeming with life could be incredibly unifying, fostering a shared sense of awe and humility. This shared experience could become the foundation for stronger international bonds, transcending differences in culture, religion, or political systems.

However, it's crucial to acknowledge the potential for the opposite effect: increased conflict and division. The discovery of extraterrestrial life could reignite existing tensions, with nations competing for access to new resources or technologies, sparking a new "space race" with potentially dangerous consequences. Differences in cultural responses to this profound discovery could lead to misunderstandings, conflict, and a widening of existing social divisions. A worst-case scenario might involve the weaponization of extraterrestrial technologies or resources, escalating existing conflicts into something far more devastating. Therefore, managing these risks would necessitate proactively developing international frameworks for conflict resolution and resource management before a discovery is made.

The importance of education and public awareness cannot be overstated. A globally coordinated educational program, designed to disseminate accurate information about extraterrestrial life and its implications, would be crucial in shaping a positive global response. This program would need to effectively communicate complex scientific concepts to diverse audiences, promoting understanding and mitigating misconceptions that could lead to fear, panic, or social unrest. Furthermore, it would have to address diverse cultural perspectives and beliefs, promoting respectful and inclusive discourse. This collaborative global educational effort might help prevent the spread of misinformation and harmful narratives, reducing the potential for fear-mongering or exploitation.

A key element in ensuring successful global cooperation would be transparency and open communication. The initial disclosure of the discovery, coupled with a commitment to open and accessible scientific data sharing, would be vital in fostering public trust and international collaboration. A coordinated, global approach to communicating the discovery to the public would mitigate the spread of misinformation and ensure a consistent, accurate narrative, reducing the risk of societal disruption. This would involve establishing clear communication channels and protocols, enabling scientists and governments to coordinate information dissemination worldwide, including establishing effective ways to address the concerns and questions of the general public.

International legal frameworks would need to be developed to govern potential interactions with extraterrestrial life. These frameworks would address a wide range of issues, including planetary protection, resource management, communication protocols, and ethical considerations. This would entail significant international negotiation and consensus-building, potentially involving the revision of existing space law and the creation of new international treaties. Furthermore, these legal frameworks would need to be adaptable, capable of adjusting to new information and unexpected challenges as we learn more about extraterrestrial life.

Ultimately, the discovery of extraterrestrial life presents humanity with a unique opportunity to transcend national boundaries and work towards a common goal. While the potential for conflict and division remains a real concern, the uniting power of a shared, extraordinary experience holds the potential to foster unprecedented global cooperation and collaboration. The success of this cooperative effort would depend on a proactive and comprehensive approach, emphasizing transparency, education, ethical considerations, and robust international frameworks for research, resource management, and conflict resolution. The challenge, however, is not merely scientific; it is fundamentally a social and political one, requiring a shift in perspective, a prioritization of global unity over national interests, and a collective commitment to a future where the discovery of extraterrestrial life serves not as a source of division, but as a catalyst for unparalleled human progress and understanding.

The discovery of extraterrestrial life would trigger an unprecedented economic upheaval, reshaping global markets and power dynamics in ways currently unimaginable. The immediate impact would likely involve a massive influx of capital into space exploration and related technologies. Governments and private companies would compete fiercely for resources and expertise, leading to a new, potentially far more intense, "space race." This competition, however, could be both beneficial and detrimental. A race focused solely on nationalistic goals might lead to a misallocation of resources and hinder true scientific progress, potentially even escalating into conflict. On the other hand, a more collaborative approach, encouraging international partnerships and sharing of knowledge and technology, could accelerate scientific breakthroughs and generate enormous economic benefits.

Imagine the potential for entirely new industries. The extraction and utilization of extraterrestrial resources, whether it's rare earth minerals from asteroids or novel energy sources from alien planets, could redefine our economic landscape. New technologies for interstellar travel, communication, and resource processing would emerge, creating entire sectors previously relegated to science fiction. The demand for skilled labor in these nascent fields would be immense, triggering massive shifts in education and workforce development. Economic inequalities might also widen dramatically, with some nations and corporations benefitting disproportionately from the exploitation of these extraterrestrial resources, leading to new global economic and social justice challenges. International cooperation would be crucial to ensure equitable distribution of benefits and prevent the emergence of a new form of economic colonialism.

The resource allocation challenges would be immense. The discovery of new resources would force a reevaluation of existing economic models and resource management strategies. The sheer volume of potential new resources could destabilize current markets, leading to price fluctuations and potentially even economic crises. International agreements would need to be established to regulate the extraction, processing, and distribution of these resources, ensuring their sustainable use and preventing conflicts over access. The establishment of international regulatory bodies, potentially within the framework of the United Nations or a newly created international organization, would be necessary to manage these challenges and prevent exploitation. This would involve complex negotiations between nations with varying economic interests and levels of technological development.

Geopolitical shifts would be profound. The balance of global power could dramatically change, with nations possessing advanced space technology and access to extraterrestrial resources gaining significant influence. The current international order, based on existing political and economic alliances, might be challenged, possibly leading to the formation of new alliances and power structures. Nations with significant space programs would have a strategic advantage, potentially leading to increased competition and even conflict. Existing international relations would be re-evaluated, requiring a complete restructuring of diplomatic strategies and policies to adapt to this new reality. The importance of diplomatic cooperation and conflict resolution mechanisms would become paramount to prevent the exploitation of the discovery for military or economic gain.

The discovery could also have profound implications for global security. The potential for extraterrestrial technology to be weaponized is a significant concern. The discovery of advanced weaponry or defensive technologies might lead to an arms race or even a direct confrontation between nations. International agreements and treaties would need to be developed to prevent the weaponization of extraterrestrial resources and ensure global security. This requires a deep consideration of ethical implications and the development of responsible guidelines for the use of these newfound technologies, potentially requiring a global reassessment of defense strategies and doctrines.

Furthermore, the economic impact extends beyond material resources. The sheer scientific and intellectual advancement spurred by the discovery would be revolutionary, transforming various fields from medicine to energy production. The resulting technological leaps could lead to breakthroughs in disease treatment, renewable energy, and materials science, positively impacting global development. This would lead to new jobs, improved living standards, and a dramatic increase in quality of life for many around the globe. However, ensuring that this advancement benefits all of humanity would require careful planning and international collaboration.

The potential for economic disparity remains a critical concern. The cost of researching and exploiting extraterrestrial resources could be prohibitively expensive, concentrating wealth and power in the hands of a few. This would necessitate the development of international mechanisms to ensure that the benefits of the discovery are shared equitably, possibly through international funding initiatives or technology transfer agreements. This equity focus must be a priority to avoid creating further economic inequalities between developed and developing nations.

Beyond the immediate economic impacts, the discovery would have long-term consequences for human society. The shift in perspective on our place in the universe, from a unique and isolated species to one among many, would force a reevaluation of our values and priorities. This could lead to significant changes in cultural norms, religious beliefs, and philosophical thought, impacting societal structures and individual lives. The ability to adapt to these changes would be crucial in ensuring societal stability and minimizing disruptions. Open dialogue, education, and intercultural understanding would be key to guiding humanity through this transformative period.

The political ramifications of the discovery would extend beyond resource management and national security. The decision-making process surrounding first contact, the sharing of information, and the regulation of interaction would require international consensus. Existing international organizations, such as the United Nations, would need to adapt their frameworks to incorporate the complexities of extraterrestrial interaction. The establishment of new international bodies specifically dedicated to managing extraterrestrial contact and the implications thereof might be necessary. International law would need to be revised to account for interactions with another intelligent species, potentially creating a new field of international space law.

The challenge of establishing international consensus on such profound matters presents significant obstacles. Differing national interests, political ideologies, and cultural perspectives would complicate the process. The establishment of transparent and equitable decision-making processes, incorporating diverse voices and viewpoints, would be essential for preventing conflict and ensuring a unified response. The potential for disagreement over policies concerning first contact, resource allocation, and the protection of extraterrestrial life would necessitate robust international mechanisms for conflict resolution.

Ultimately, the economic and political repercussions of discovering extraterrestrial life would be inextricably linked. Economic competition could exacerbate existing geopolitical tensions, while international cooperation could foster economic growth and global prosperity. The success of navigating this complex landscape would depend on a proactive and comprehensive approach, prioritizing international collaboration, ethical considerations, and equitable resource allocation. The discovery of extraterrestrial life would be a watershed moment in human history, demanding a fundamental shift in our perspective and a collective commitment to a future where cooperation and understanding prevail. The challenge is not just to understand the universe but also to understand and manage ourselves as a species on the threshold of a transformative encounter. The stakes are incredibly high, and the potential for both extraordinary advancement and catastrophic failure are equally profound.

The discovery of extraterrestrial life would undoubtedly trigger a seismic shift in human culture and religious beliefs, a transformation potentially more profound than any event in recorded history. For millennia, humanity's understanding of its place in the cosmos has been shaped by religious and cultural narratives. These narratives, often intertwined with mythology and folklore, have provided frameworks for understanding our origins, purpose, and relationship with the universe. The sudden revelation of other intelligent life forms would challenge, reinterpret, or even obliterate many of these foundational beliefs.

Consider the implications for monotheistic religions, where the concept of a single, all-powerful creator god is central. The existence of extraterrestrial life could raise questions about the nature of creation, the uniqueness of humanity, and the divine plan. Some might interpret this discovery as evidence supporting their existing beliefs, viewing extraterrestrial life as a part of a larger, divinely orchestrated cosmic order. Others might find their faith shaken, grappling with the implications of a universe potentially teeming with life forms created outside their theological frameworks. The ensuing theological debates and reinterpretations could be widespread and intense, leading to significant internal divisions within religious communities. The potential for both strengthened faith and widespread apostasy is equally plausible.

The impact on polytheistic and animistic belief systems could be similarly complex. These systems, often characterized by a diverse pantheon of deities and spirits inhabiting the natural world, may adapt more readily to the concept of extraterrestrial life. The inclusion of new deities or spirits representing these other life forms could seamlessly integrate into existing cosmologies. However, the discovery might also lead to a reevaluation of the power dynamics within these pantheons, sparking internal conflict and reorganization within the religious structures. The potential for syncretism – the blending of different religious traditions – is high, creating new, hybrid belief systems that incorporate elements of both terrestrial and extraterrestrial religions.

Beyond organized religion, the discovery would have a profound impact on secular worldviews and cultural narratives. Humanity's self-perception as the pinnacle of evolution, the sole possessors of intelligence and consciousness in the universe, would be fundamentally challenged. The anthropocentric view, which places humankind at the center of the cosmos, would be irrevocably shattered. This would lead to a reevaluation of human values, ethics, and social structures. Questions of our responsibility towards extraterrestrial life, the nature of interspecies communication, and the potential for conflict would need to be addressed.

The arts and humanities would be dramatically transformed. Literature, music, film, and visual arts would explore the new possibilities presented by extraterrestrial contact, leading to new forms of expression and aesthetic interpretations. Science fiction, already exploring these themes, would experience a surge in both popularity and influence, potentially becoming a key tool for understanding and navigating this new reality. The very definition of "humanity" would undergo a period of intense reevaluation, potentially leading to a more inclusive and cosmopolitan self-understanding.

The impact on scientific endeavors would also be significant. Existing scientific theories and methodologies would need to be revisited and updated to incorporate the findings of extraterrestrial life studies. New fields of research would emerge, focusing on astrobiology, xenolinguistics, and interspecies communication, attracting significant funding and talent. International collaborations would become even more critical, necessitating the establishment of new research institutions and collaborations to address the unprecedented scientific challenges that would arise.

Education systems would undergo profound changes as well. The curriculum would need to be reformed to incorporate the new scientific understanding of the universe and humanity's place within it. The teaching of history, philosophy, and religious studies would be transformed, requiring critical analyses of our previous assumptions and biases. New educational materials would be developed, bridging the gap between scientific findings and their cultural and societal implications. An emphasis on intercultural understanding and responsible ethical decision-making would be crucial in preparing future generations for a world sharing the cosmos with other intelligent life forms.

The discovery could lead to both unprecedented cooperation and heightened conflict. The shared challenge of understanding and interacting with extraterrestrial life could foster a sense of global unity and shared purpose, transcending national borders and political divisions. However, the competition for resources, technology, and influence could also intensify, potentially leading to new forms of conflict and competition. The potential for misunderstanding, miscommunication, and even warfare would necessitate careful planning and international cooperation.

The role of mass media in shaping public perception would be pivotal. The media's responsibility in disseminating accurate information, promoting critical thinking, and fostering constructive dialogue would be paramount.

The challenge would lie in mitigating misinformation and harmful narratives, while simultaneously engaging the public in a way that fosters understanding and acceptance of the profound implications of this discovery.

Furthermore, the economic and political implications discussed previously would inextricably influence cultural and religious shifts. Economic inequality resulting from the unequal distribution of extraterrestrial resources could exacerbate social divisions and religious tensions. Political power struggles over the control of space-related technologies and resources could fuel nationalist sentiments and hinder international cooperation. The management of these economic and political factors would be critical in ensuring a smooth transition to a world where humanity shares its cosmic neighborhood with other intelligent life forms.

Ultimately, the societal impact of discovering extraterrestrial life would be an ongoing and evolving process. The initial shock and awe would gradually give way to a period of adaptation, interpretation, and integration. The way humanity responds to this discovery – whether through cooperation, conflict, or a combination of both – would define the future trajectory of our civilization and our relationship with the universe. The challenge lies in navigating this transformative period with wisdom, compassion, and a commitment to a future where understanding and peaceful coexistence prevail, both among ourselves and with the inhabitants of the cosmos. The discovery promises not just a new era of scientific advancement but a fundamental reassessment of our identity, beliefs, and our place in the vast, unexplored expanse of the universe.

The potential discovery of extraterrestrial life necessitates a proactive and comprehensive approach to public education and awareness. This is not merely a matter of disseminating scientific findings; it's about fostering a societal understanding that can navigate the profound cultural, religious, philosophical, and political shifts such a discovery would inevitably trigger. The sheer magnitude of this event demands a carefully planned and executed educational strategy that reaches all segments of society, regardless of age, background, or belief system.

One crucial aspect of this educational endeavor is the reform of existing curricula across various disciplines. Science education, naturally, requires a significant overhaul. Astrobiology, previously a niche field, would need to become a cornerstone of scientific literacy. Students should not simply learn about the possibility of extraterrestrial life; they should understand the underlying scientific principles, the methodologies used to search for it, and the potential implications of various discoveries. This means incorporating the latest research in planetary science, exobiology, and SETI (Search for Extraterrestrial Intelligence) into primary, secondary, and tertiary education. Curriculum designers should move beyond the traditional textbook approach, embracing interactive learning methods such as simulations, virtual reality experiences, and citizen science projects that engage students directly in the scientific process.

Beyond science, the humanities and social sciences also require a significant update. History curricula could explore the cultural and religious responses to past astronomical discoveries and incorporate comparative studies of different societies' beliefs about the cosmos. Philosophy courses should delve into the ethical implications of contact, the challenges of intercultural communication, and the potential for both cooperation and conflict. Religious studies should engage in open and critical discussions of how various faith traditions might interpret the existence of extraterrestrial life, promoting understanding and tolerance amongst diverse perspectives. The integration of these subjects ensures a holistic approach, equipping individuals with the tools to critically analyze the societal implications of such a transformative event.

Furthermore, the arts play a vital role in preparing society for this potential paradigm shift. The exploration of extraterrestrial themes in literature, film, music, and visual arts can help demystify the topic, fostering empathy and understanding. By engaging with fictional narratives that grapple with the challenges and opportunities of contact, individuals can prepare themselves for a reality that may soon become tangible. Encouraging creative expression around the subject can also foster critical thinking and open dialogue about the complex societal issues involved.

Beyond formal education, public awareness campaigns are equally crucial. These campaigns should be multifaceted, targeting diverse audiences through various media channels. Clear, concise, and accurate information, presented in an accessible and engaging manner, is crucial to counter the spread of misinformation and speculation. Scientific institutions, governments, and media organizations must work together to develop and disseminate these campaigns, utilizing various forms of media, from social media platforms to television documentaries and public lectures. The emphasis should be on fostering open dialogue and promoting critical thinking.

It's crucial to avoid a top-down, didactic approach to public education. Instead, a participatory model that actively engages the public is essential. Citizen science projects offer a powerful avenue for public engagement, allowing individuals to contribute directly to the search for extraterrestrial life and the understanding of its implications. This hands-on approach can increase public interest, scientific literacy, and a sense of shared responsibility in addressing the challenges this discovery would bring. Public forums, debates, and interactive discussions can facilitate open conversations, promoting critical thinking and fostering a sense of community around the topic.

The role of museums and science centers should also be emphasized. These institutions can provide interactive exhibits and educational programs that engage the public with the science behind the search for extraterrestrial life and the potential societal implications of its discovery. They can showcase the latest research, host lectures by leading scientists, and create spaces for open dialogue and discussion. These institutions should not just present facts but also engage visitors emotionally, fostering a sense of wonder and encouraging deeper reflection.

Media literacy is another critical component of effective public education. The dissemination of accurate and reliable information is paramount. Combating misinformation and conspiracy theories that may arise surrounding a discovery is essential. The media has a critical role to play in verifying information, providing a platform for scientific experts, and educating the public about the complexities of the issue. Encouraging media literacy helps individuals develop critical thinking skills, allowing them to discern credible sources from unreliable ones.

In addition to addressing the general public, specific attention must be given to vulnerable and marginalized communities. The societal impact of such a discovery will not be uniformly experienced across society. It is important to ensure that education and awareness campaigns are culturally sensitive and accessible to everyone, regardless of socioeconomic status, language, or disability. A participatory approach that involves these communities in the shaping and delivery of educational programs is vital for fostering inclusion and ensuring that all voices are heard.

Finally, it's essential to acknowledge the uncertainties and ambiguities inherent in the quest for extraterrestrial life. The educational effort must not present a simplistic or overly optimistic narrative. It should prepare society for a range of potential outcomes, including the possibility of non-contact, the potential for misunderstanding, and even the possibility of conflict. Openly discussing these uncertainties fosters resilience and critical thinking, helping society navigate the complexities of such a transformative event. The goal is not to create a uniform response but to cultivate a critical and adaptable society capable of grappling with the profound implications of a discovery that could reshape our understanding of the universe and our place within it. A proactive and thoughtful approach to education and public awareness is therefore not merely desirable—it is essential for the future of humanity.

The potential discovery of extraterrestrial life presents a unique challenge: managing the intense public interest and the inevitable flood of misinformation that would accompany such a monumental announcement. Responsible communication, therefore, becomes paramount, not just in disseminating scientific findings, but in shaping a societal response that is both informed and constructive. This requires a multi-pronged strategy that encompasses proactive public education, rigorous fact-checking, and a commitment to transparent communication from scientific institutions, governments, and media outlets.

One crucial aspect of this strategy is establishing clear channels for the dissemination of information. A dedicated, centralized source of information, potentially a government-backed agency or an international consortium of scientific organizations, could serve as the primary point of contact for the public. This agency could be responsible for verifying the authenticity of any claims, providing regular updates on the ongoing investigation, and offering interpretations of the scientific findings in an accessible and non-technical manner. This would help to prevent the spread of rumors and speculation, replacing them with reliable, verifiable information.

The language used in communicating this information is equally critical. Scientific jargon must be avoided, and complex concepts should be explained using clear, concise, and readily understandable language. Utilizing diverse communication formats – infographics, videos, interactive online resources – can broaden the reach and accessibility of information, catering to different learning styles and preferences. Moreover, the communication should be culturally sensitive, taking into account the diverse range of beliefs and cultural interpretations that exist within different communities. Translation into multiple languages is essential to ensure that information reaches a global audience.

Transparency is key to building trust and combating misinformation. The public needs to understand the scientific process, the limitations of the data, and the uncertainties inherent in such a discovery. Openly acknowledging these uncertainties and limitations prevents the creation of unrealistic expectations and helps to foster a more nuanced understanding of the implications of extraterrestrial contact. The communication strategy should not shy away from the complexities of the situation, but should, instead, embrace them, fostering a sense of shared exploration and understanding.

Combating misinformation requires a proactive approach. Dedicated fact-checking initiatives, supported by leading scientists and communication experts, could actively monitor online and offline discussions, identifying and debunking false or misleading information. These initiatives should not simply flag misinformation; they should provide accurate counter-narratives that address the core concerns and misconceptions fueling the spread of falsehoods. Partnerships with social media platforms and other media outlets could be instrumental in ensuring the rapid dissemination of these counter-narratives.

Another crucial element in managing expectations is the framing of the narrative. The communication surrounding any potential discovery needs to avoid hyperbole and sensationalism. It is vital to present the information in a measured and balanced manner, highlighting the scientific evidence and avoiding speculation or conjecture. Framing the narrative around scientific inquiry, rather than sensational claims, helps to foster a more grounded and informed understanding. The language used should emphasize the wonder and potential of discovery, while also acknowledging the potential challenges and implications.

Furthermore, engaging experts from diverse fields is crucial for a comprehensive and nuanced understanding of the societal implications. Scientists, theologians, philosophers, anthropologists, sociologists, and political scientists, among others, can offer unique perspectives on how such a discovery could reshape our worldview, our societal structures, and our very understanding of humanity's place in the universe. Including these diverse perspectives in the communication strategy ensures a holistic approach, addressing the potential impact across different sectors of society.

The role of media literacy education cannot be overstated. A population equipped with the skills to critically evaluate information is better equipped to resist misinformation and to engage with scientific findings in a responsible manner. This necessitates a nationwide effort to improve media literacy education at all levels, from primary school to higher education. The curriculum should equip students with the tools to identify credible sources, distinguish facts from opinions, and recognize biased or misleading narratives.

Addressing potential psychological and societal impacts is also crucial. The discovery of extraterrestrial life could trigger profound psychological responses in individuals and communities. A well-planned communication strategy should anticipate these reactions and develop strategies to manage them. This might involve providing access to mental health support, fostering community dialogue, and facilitating opportunities for individuals to process and share their reactions in a safe and supportive environment.

The ethical implications of contact require careful consideration. Questions of how to interact with an extraterrestrial civilization, how to manage potential resource conflicts, and how to ensure the protection of both humanity and any extraterrestrial life encountered must be addressed openly and honestly. A proactive ethical framework, developed in consultation with experts from various fields, should be established to guide decision-making and to ensure that any interaction is conducted in a way that respects the potential rights and interests of all involved parties.

The long-term societal impact of contact could extend beyond immediate reactions. Economic shifts, political realignments, and potential changes in religious and philosophical beliefs are all possibilities. A robust communication strategy must consider these potential long-term consequences and prepare society for a gradual and potentially transformative adaptation. This requires ongoing engagement and dialogue, building a resilient and adaptable society capable of navigating a future shaped by the discovery of life beyond Earth.

Finally, international cooperation is essential for managing the global implications of such a discovery. An extraterrestrial discovery is not a national issue; it is a global one. International collaborations, involving scientific organizations, governments, and other relevant stakeholders, are necessary to ensure a coordinated and unified response. This requires establishing clear communication protocols, harmonizing data sharing procedures, and promoting a collaborative approach to understanding and managing the societal implications of contact. A unified, global approach will foster trust and transparency, counteracting the spread of misinformation and contributing to a more informed and responsible societal response. The quest for extraterrestrial life is a collective endeavor, and managing the impact of such a discovery must reflect this fundamental truth.

| 15 |

A Universe of Possibilities

Having explored the diverse tapestry of life on Earth, from the extremophiles thriving in seemingly impossible environments to the complex ecosystems that shape our planet, we've established a powerful baseline for considering the potential for life beyond our world. This book hasn't offered definitive proof of extraterrestrial life, nor has it aimed to. Instead, it's presented a compelling case built upon reasoned speculation and the meticulous analysis of available data, weaving together scientific findings with the intriguing narratives surrounding reported UFO encounters. The goal has been to stimulate thought, to encourage a deeper understanding of the vast possibilities inherent in the cosmos, and to prepare for a future where the discovery of extraterrestrial life might no longer be a matter of science fiction, but of scientific fact.

One of the central arguments throughout this exploration has been the sheer improbability of Earth being unique. The incredible diversity of life found on our planet, its capacity to adapt and thrive in even the most extreme conditions, speaks volumes about the resilience and adaptability of life itself. Consider the extremophiles – organisms that flourish in environments previously thought inhospitable, from the scalding vents of hydrothermal systems to the frigid depths of subglacial lakes. These organisms challenge our narrow definitions of "habitable," suggesting that life might exist in places we haven't even begun to imagine. The presence of liquid water, once considered a prerequisite for life, is increasingly viewed as only one element within a complex set of potentially suitable conditions.

This broadening of our understanding of habitability dramatically expands the number of potential locations for life beyond Earth. The moons of our solar system, such as Europa and Enceladus, with their subsurface oceans, are prime candidates for harboring life. Exoplanet research, too, has exploded in recent years, revealing thousands of planets orbiting other stars. While we're still in the early stages of characterizing these exoplanets, the sheer number already discovered suggests that Earth-like planets – or planets with conditions potentially suitable for life – are far more common than once thought. The possibility of life arising independently on multiple planets within our own galaxy, let alone across the vast expanse of the universe, becomes increasingly plausible.

The reported accounts of UFO encounters, while often shrouded in controversy and lacking definitive scientific evidence, raise fascinating questions about the possibility of extraterrestrial contact. These accounts, while not sufficient to confirm the existence of extraterrestrial intelligence, provide a compelling context for examining the potential implications of such contact. The common themes that emerge from many of these reports – encounters with advanced technology, unusual physiological effects, and experiences suggesting a sophisticated understanding of human psychology – suggest a need for further investigation, not dismissal.

This book has emphasized the importance of responsible speculation. While scientific rigor must remain paramount, dismissing potential evidence based solely on a lack of definitive proof is shortsighted. The exploration of extraterrestrial life requires a willingness to embrace uncertainty, to carefully weigh available evidence, and to remain open to the possibility of unexpected discoveries. The scientific method itself relies on hypothesis testing, and the hypothesis of extraterrestrial life demands investigation.

Furthermore, the potential discovery of extraterrestrial life necessitates careful consideration of its societal implications. The management of public perception, the dissemination of information in a clear and accessible manner, and the fostering of a globally coordinated response are all crucial considerations. This book has stressed the importance of proactive communication strategies, including transparent and accountable government and scientific institutions, to counteract the inevitable spread of misinformation that would accompany such an announcement.

The ethical dilemmas presented by potential contact are equally significant. How should humanity approach the discovery of another intelligent species? What protocols should be followed to ensure the safety and well-being of both parties? These are questions that require careful consideration by scientists, ethicists, policymakers, and the global community as a whole. A unified, international approach to these ethical considerations is paramount to ensure responsible and respectful engagement.

The long-term societal consequences of such a discovery are likely to be profound and far-reaching. The impact on our scientific understanding of the universe, our philosophical perspectives on humanity's place within it, and even our religious and spiritual beliefs could be transformative. Preparing for these potential changes requires a proactive approach, encompassing education, open dialogue, and the development of adaptable societal structures that can accommodate a future vastly different from the one we know today.

The exploration of extraterrestrial life is not simply a scientific pursuit; it is a journey of self-discovery. By examining the possibility of life beyond Earth, we gain a deeper understanding of the conditions necessary for life, the resilience of life itself, and the potential for the universe to harbor an abundance of life forms we can only begin to imagine. This exploration fosters a sense of wonder, reminding us of the vastness of the cosmos and our own place within it. It encourages critical thinking, emphasizing the importance of evidence-based reasoning and the necessity of challenging our assumptions. Ultimately, it invites us to consider a future where our place in the universe is not isolated, but interconnected, enriching our understanding of ourselves and our place in the grand scheme of existence.

The quest for extraterrestrial life continues. While definitive proof remains elusive, the scientific evidence, coupled with the intriguing accounts of UFO encounters and the increasing understanding of the universe's potential for life, create a compelling case for continued investigation and informed speculation. This book serves as a testament to the power of scientific curiosity, the importance of responsible speculation, and the need to prepare for a future that might be far more extraordinary than we have ever imagined. The universe may indeed teem with life, a universe of possibilities waiting to be unveiled. This exploration, while filled with uncertainties, is a testament to the enduring human spirit of exploration and discovery, a quest that will continue for generations to come.

Our journey through the cosmos has only just begun. The universe, with its vastness and potential for life beyond our wildest dreams, presents an endless frontier for exploration, scientific discovery, and the expansion of our understanding of life itself. The implications of discovering life beyond Earth are immense, challenging our very understanding of ourselves, our place in the universe, and the potential future of humanity. This exploration transcends mere scientific inquiry; it is a testament to the boundless curiosity that drives us to push the limits of knowledge, to unravel the mysteries of the universe, and to strive for a future where humanity's place in the cosmos is not one of isolation, but of interconnectedness and understanding.

The journey we have undertaken in this book has not been one of definitive answers but of reasoned exploration, a journey that balances scientific rigor with the inherent wonder of the cosmos. We've explored the incredible biodiversity of our planet, demonstrating the adaptability of life and expanding the definition of habitability to encompass environments once thought impossible. We've delved into the fascinating, albeit often controversial, narratives surrounding UFO encounters, viewing them as a springboard for discussion and further scientific investigation. And we've examined the potential societal and ethical implications of encountering extraterrestrial life, emphasizing the importance of responsible communication, international collaboration, and a proactive approach to managing the profound implications of such a discovery.

As we conclude this journey, the overarching message remains one of possibility and the enduring human spirit of exploration. The search for extraterrestrial life is a testament to our innate curiosity, our desire to understand our place in the universe, and our capacity to adapt and evolve in the face of the unknown. Whether or not we ultimately find definitive proof of life beyond Earth, the quest itself has already expanded our understanding of the universe, of life itself, and of humanity's place within this magnificent and mysterious cosmos. The journey continues. The exploration never ends. The universe, with all its wonders and mysteries, awaits.

The preceding chapters have laid out a compelling case for the possibility of extraterrestrial life, weaving together scientific evidence with the intriguing narratives surrounding UFO encounters. However, many questions remain unanswered, and the path forward demands a multidisciplinary approach, integrating various scientific disciplines and fostering international collaboration. This section outlines some of the most pressing unresolved questions and proposes future research directions that could significantly advance our understanding of the universe and our place within it.

One of the most fundamental unresolved questions is the origin of life itself. While we understand the basic building blocks of life – DNA, RNA, and proteins – the precise process by which these molecules assembled to form the first self-replicating entities remains a mystery. Understanding abiogenesis, the transition from non-living matter to living organisms, is crucial to evaluating the likelihood of life arising elsewhere in the universe. Further research into the chemistry of early Earth, coupled with experiments simulating prebiotic conditions, may shed light on this crucial process. The discovery of alternative biochemistries, beyond the carbon-based life we know, would revolutionize our understanding of life's potential and expand the search for life beyond Earth to environments and chemical compositions previously deemed inhospitable.

The search for extraterrestrial life is fundamentally limited by our current technological capabilities. While ground-based and space-based telescopes have revolutionized our ability to detect exoplanets, characterizing their atmospheres and identifying biosignatures remains a significant challenge. Next-generation telescopes, such as the Extremely Large Telescope (ELT) and the James Webb Space Telescope (JWST), will provide unprecedented sensitivity and resolution, enabling the detailed study of exoplanet atmospheres and the search for biosignatures such as oxygen, methane, and water vapor in specific ratios indicative of biological activity. Furthermore, advanced spectroscopic techniques, coupled with machine learning algorithms, will be vital in analyzing vast amounts of data and identifying subtle patterns indicative of life. The development of more sensitive instrumentation and innovative data analysis methods is crucial for expanding the range of detectable biosignatures and improving our ability to differentiate between biological and geological processes.

Beyond our immediate solar system, the sheer distances involved in interstellar travel present a significant obstacle to direct exploration. However, advances in propulsion technology, such as nuclear fusion propulsion or advanced laser-based systems, could potentially reduce travel times to nearby star systems, making interstellar missions feasible within the foreseeable future. The development of autonomous probes capable of long-duration missions and equipped with advanced scientific instrumentation is also crucial. Such probes could collect samples from exoplanets and their moons, conduct in-situ analyses, and potentially return samples to Earth for detailed study. This will require substantial investment in advanced robotics, artificial intelligence, and miniaturized instrumentation.

The question of the prevalence of life beyond Earth is intertwined with our understanding of planetary formation and evolution. The study of planetary systems around other stars, particularly those with planets in the habitable zone, is crucial to understanding the conditions that favor the emergence and sustenance of life. Detailed characterization of exoplanet atmospheres, coupled with models of planetary formation and evolution, will help refine our understanding of habitability and identify promising targets for future searches for life. Furthermore, comparative planetology, which involves comparing the characteristics of various planets within our solar system and beyond, can help identify common features and patterns that may be associated with the emergence and evolution of life.

The search for extraterrestrial intelligence (SETI) continues to be a challenging but potentially transformative endeavor. While radio astronomy remains a key tool in the search for extraterrestrial signals, the development of more advanced search techniques, including the monitoring of other electromagnetic frequencies and the search for other forms of communication, is crucial. The development of powerful artificial intelligence and machine learning algorithms will be essential to sift through the vast amounts of data generated by these searches, identifying potential signals that may have been missed by previous methods. The integration of SETI research with astrobiological studies is also essential, to coordinate efforts and to develop a more comprehensive understanding of the potential for life and intelligence beyond Earth.

The societal implications of discovering extraterrestrial life cannot be overstated. Preparing for such an event requires careful planning and international collaboration. The development of proactive communication strategies, involving scientists, policymakers, and the public, is essential to ensure the responsible dissemination of information and to prevent the spread of misinformation. Establishing international protocols for responding to potential extraterrestrial contact, addressing both scientific and ethical considerations, is crucial for ensuring a unified and responsible global response.

In addition to the scientific and technological challenges, the ethical considerations surrounding the search for and potential discovery of extraterrestrial life are profound.

Questions regarding the potential impact on religious and spiritual beliefs, the possible exploitation of extraterrestrial life or resources, and the ethical implications of interstellar travel demand careful consideration by ethicists, policymakers, and the global community. The establishment of international ethical guidelines, informed by scientific understanding and guided by ethical principles, is crucial for navigating the potential challenges and opportunities presented by extraterrestrial contact. Open and transparent discussions about these complex ethical issues are vital for ensuring a responsible approach to the search for and potential encounter with extraterrestrial life.

The search for extraterrestrial life is a global endeavor, requiring the collaboration of scientists, engineers, policymakers, and the public. International collaborations, sharing data and resources across national boundaries, are essential to accelerate progress in this field. The establishment of global networks for data sharing and collaborative research will facilitate the advancement of knowledge and the development of new technologies. Furthermore, fostering education and public engagement through outreach programs and educational initiatives is crucial for increasing public understanding of the search for extraterrestrial life and promoting global support for these important endeavors. A unified, internationally coordinated effort is paramount to successfully navigate the challenges and opportunities presented by this grand scientific pursuit.

The quest for extraterrestrial life is a long-term endeavor that will likely span multiple generations. Maintaining consistent funding and support for research, education, and international collaboration is crucial for sustained progress. The exploration of extraterrestrial life transcends mere scientific inquiry; it is a fundamental human quest, reflecting our innate curiosity, our desire to understand our place in the universe, and our capacity for exploration and discovery. The journey continues, and the possibilities are limitless. The universe awaits.

The journey to understand our cosmic neighborhood and the potential for life beyond Earth is far from over. The preceding chapters have highlighted the compelling evidence suggesting the possibility of extraterrestrial life, yet numerous questions remain unanswered. The path forward requires not only continued exploration but a paradigm shift in our approach – one that embraces interdisciplinary collaboration, innovative technologies, and a long-term commitment to unraveling the universe's deepest mysteries. This unwavering pursuit is not merely a scientific endeavor; it is a testament to humanity's inherent curiosity, our drive to understand our place in the cosmos, and our potential to uncover profound truths about existence itself.

One of the most crucial areas requiring sustained investigation is the origin of life. While we've made significant strides in understanding the building blocks of life on Earth, the exact process by which non-living matter transitioned into self-replicating organisms—abiogenesis—remains elusive. This puzzle extends beyond our planet; comprehending abiogenesis is essential for assessing the probability of life's emergence elsewhere in the universe. Future research must focus on detailed studies of the early Earth's environment, employing advanced simulations to recreate prebiotic conditions and analyze the chemical reactions that might have led to the first living cells. Moreover, exploring the possibility of alternative biochemistries, beyond the carbon-based life forms we know, is crucial. Such discoveries would revolutionize our understanding of life's potential and significantly broaden the scope of our search for extraterrestrial life, encompassing environments and chemical compositions previously deemed inhospitable. The discovery of silicon-based life, for instance, or life forms utilizing radically different solvents than water, would force a complete re-evaluation of our assumptions about habitability.

Technological advancements are paramount in furthering our understanding of the cosmos and the search for life beyond Earth. While current telescopes have provided unprecedented insights into exoplanets, characterizing their atmospheres and identifying definitive biosignatures remains a challenge. Future generations of telescopes, like the Extremely Large Telescope (ELT) and the Nancy Grace Roman Space Telescope, will offer exponentially improved sensitivity and resolution, enabling detailed atmospheric analyses and more precise detection of biosignatures like oxygen, methane, and water vapor—telltale signs of biological activity. However, simply collecting data is insufficient; sophisticated data analysis techniques, including machine learning algorithms, are crucial for processing the vast amounts of information gathered, identifying subtle patterns, and discerning biosignatures from geological or other non-biological processes. The development of novel spectroscopic techniques and algorithms that can analyze light from exoplanets with unprecedented accuracy will be essential in confirming the presence of life-related chemicals.

Interstellar travel remains a significant hurdle, but persistent research into propulsion systems is slowly chipping away at this challenge. Nuclear fusion propulsion and advanced laser-based systems offer potential pathways to significantly reduce travel times to nearby star systems, making interstellar missions increasingly feasible. Alongside propulsion breakthroughs, the development of autonomous probes capable of long-duration missions and equipped with advanced scientific instrumentation is equally vital. These probes could collect samples from exoplanets and their moons, conduct in-situ analyses, and potentially return samples to Earth for more thorough examination. This endeavor will require substantial investment in robotics, artificial intelligence, and miniaturization of scientific equipment to create probes capable of functioning reliably in the harsh environments of distant worlds. The development of robust, self-repairing systems will be key to the success of such long-term missions.

The question of the prevalence of life beyond Earth is closely intertwined with our comprehension of planetary formation and evolution. The study of planetary systems around other stars, especially those with planets in the habitable zone, is crucial for understanding the conditions conducive to life's emergence and sustainability. Through detailed atmospheric characterization of exoplanets, coupled with sophisticated models of planetary formation and evolution, we can refine our understanding of habitability and identify promising targets for future searches for life. This effort must go hand-in-hand with comparative planetology, meticulously comparing the characteristics of planets within our solar system and beyond, to uncover common features and patterns that could be indicative of life's emergence and evolution. The comparison of diverse planetary environments might reveal unexpected correlations and lead to novel theories about the emergence of life.

The search for extraterrestrial intelligence (SETI) continues to be a challenging but potentially transformative pursuit. While radio astronomy remains a pivotal tool, the development of more advanced search techniques, encompassing other electromagnetic frequencies and potentially novel communication methods, is essential. Powerful artificial intelligence and machine learning algorithms are indispensable for sifting through the immense quantities of data generated by SETI searches, identifying potential signals that might have been overlooked by previous methods. The integration of SETI research with astrobiological studies is equally important, facilitating a more comprehensive understanding of the potential for life and intelligence beyond Earth. A coordinated effort between these fields could lead to more targeted searches and a more profound understanding of the cosmic landscape.

The societal implications of discovering extraterrestrial life are profound and demand careful consideration. Preparing for such an event requires proactive planning and international collaboration. The development of clear and effective communication strategies, engaging scientists, policymakers, and the public, is crucial for responsible dissemination of information and preventing the spread of misinformation. International protocols for responding to potential extraterrestrial contact must be established, addressing both scientific and ethical considerations to ensure a unified and responsible global response. Such protocols should aim to maximize benefits and minimize potential risks, both to humanity and any extraterrestrial life encountered.

Beyond the scientific and technological hurdles, the ethical considerations are equally profound. The potential impact on religious and spiritual beliefs, the possible exploitation of extraterrestrial life or resources, and the ethical implications of interstellar travel all require careful consideration by ethicists, policymakers, and the global community. The establishment of international ethical guidelines, grounded in scientific knowledge and ethical principles, is critical for navigating the potential challenges and opportunities presented by extraterrestrial contact. Open and transparent dialogue about these complex ethical issues is crucial for a responsible approach to the search for and potential encounter with extraterrestrial life. This ethical framework must ensure that any interaction with extraterrestrial life is conducted with respect, consideration, and a commitment to the long-term well-being of all involved.

The search for extraterrestrial life demands a global collaboration involving scientists, engineers, policymakers, and the public. International collaborations, sharing data and resources across national boundaries, are essential for accelerating progress in this field. The establishment of global networks for data sharing and collaborative research will facilitate the advancement of knowledge and the development of new technologies. Furthermore, fostering education and public engagement through outreach programs and educational initiatives is vital for increasing public understanding of the search for extraterrestrial life and promoting global support for these important endeavors. A unified, internationally coordinated effort is paramount to successfully address the challenges and exploit the opportunities presented by this grand scientific pursuit.

The quest for extraterrestrial life is a long-term endeavor, spanning multiple generations. Maintaining consistent funding and support for research, education, and international collaboration is crucial for sustained progress. The exploration of extraterrestrial life transcends mere scientific inquiry; it's a fundamental human quest, reflecting our innate curiosity, our desire to understand our place in the universe, and our capacity for exploration and discovery. The journey continues, and the possibilities, as vast as the universe itself, remain limitless. The exploration of the cosmos and the search for life beyond Earth is not just a scientific pursuit; it is a defining characteristic of humanity's ambition and an enduring testament to our quest for knowledge and understanding. The universe awaits, and the future of exploration promises untold discoveries.

The discovery of extraterrestrial life, even microbial, would irrevocably alter our understanding of our place in the cosmos. For millennia, humanity has pondered its uniqueness, its position as the sole intelligent species in a vast and seemingly indifferent universe. The Copernican revolution shifted our understanding of Earth's place in the solar system, demoting our planet from the center of creation to a relatively insignificant world orbiting a rather ordinary star. The discovery of exoplanets further broadened our perspective, revealing a universe teeming with planetary systems, many potentially harboring environments suitable for life. Yet, the confirmation of extraterrestrial life would represent a paradigm shift of even greater magnitude, shattering the anthropocentric worldview that has dominated human thought for centuries. It would force us to confront our assumptions about life's uniqueness and the potential for intelligence beyond our own. The implications would ripple through every aspect of human society, from religion and philosophy to science and politics.

Religious beliefs, deeply intertwined with humanity's understanding of its origin and purpose, would undoubtedly be challenged. Many faiths posit humanity as a unique creation, endowed with a special relationship with a divine being. The discovery of extraterrestrial life, particularly intelligent life, could lead to a reassessment of these beliefs, raising profound questions about the nature of creation, the role of humanity in the universe, and the existence and nature of any divine power. Some religious interpretations might embrace the discovery, viewing extraterrestrial life as further evidence of a vast and mysterious creator. Others might struggle to reconcile the discovery with established dogma, potentially leading to theological debates and shifts in religious thought. The integration of extraterrestrial life into our understanding of the divine will require a critical re-evaluation of existing religious doctrines, potentially leading to the emergence of new theological interpretations or a reevaluation of existing ones.

Philosophically, the discovery would prompt a reevaluation of fundamental concepts such as consciousness, intelligence, and the nature of existence. Are there universal principles governing the emergence and evolution of life? Does consciousness arise inevitably from certain levels of complexity, or is it a unique characteristic of certain life forms? If extraterrestrial life displays different forms of intelligence or consciousness, would it broaden our understanding of these concepts beyond the limitations of our current human perspective? The potential for radically different forms of life, with unique biochemical pathways, communication systems, and societal structures, challenges the very foundations of our philosophical inquiry into the meaning and purpose of existence. It would open up new avenues of philosophical inquiry, forcing us to re-examine the nature of reality, consciousness, and the place of humanity within a much wider cosmic context. The very definition of "humanity" would be questioned, potentially leading to a reassessment of our ethical responsibilities, both towards any discovered extraterrestrial life and towards each other.

The societal impact of such a discovery would be equally profound. International relations would undergo a dramatic transformation as nations collaborate to address the implications of extraterrestrial contact. The sharing of information, technology, and resources would likely become essential for understanding and responding to this momentous event. Competition and conflict, however, are not unlikely, potentially driven by disputes over resources, territorial claims, or technological dominance in the exploration and exploitation of extraterrestrial environments or resources. A new era of international cooperation, potentially exceeding current levels of collaboration in space exploration, would be crucial. International treaties and ethical guidelines would need to be developed to govern our interactions with extraterrestrial life, ensuring the responsible and ethical exploration of any discovered worlds. This process necessitates a global effort, requiring significant diplomatic and political engagement to achieve consensus and avoid conflicts. The governance structure for this new phase of cosmic exploration would need to be established, taking into account the potential challenges and opportunities.

The economic consequences could also be transformative. New industries focused on space exploration, extraterrestrial resource extraction, and interstellar communication would emerge, potentially reshaping global economies. The potential discovery of valuable resources on other planets or celestial bodies could lead to economic booms, but also to heightened competition and potential conflicts over access to these resources. Investment in space-related technologies would receive a massive influx of resources and interest, potentially diverting funds from other areas of research and development. However, the economic impact wouldn't be limited to tangible resources. The development of new technologies and scientific knowledge through interactions with extraterrestrial civilizations could lead to breakthroughs in various fields, transforming various aspects of human civilization, from medicine and engineering to communication and transportation. This could trigger a technological singularity, an era of rapid and unforeseen technological development. This shift could have an unforeseen and profound effect on global economics.

Beyond the immediate practical and economic implications, the discovery of extraterrestrial life would have a profound psychological impact on humanity. The potential for encountering life forms radically different from ourselves could challenge our anthropocentric views, altering our self-perception and broadening our understanding of the diversity of life in the universe. This encounter could lead to a sense of wonder and awe, inspiring new levels of scientific and artistic creativity. Conversely, it could also evoke fear and anxiety, prompting a reevaluation of humanity's place and purpose in the cosmos. Such a discovery would force humanity to confront its own existential fears and insecurities, triggering profound psychological shifts. The psychological impact might manifest differently across cultures, with some embracing the discovery as a unifying force, while others react with fear and uncertainty. This requires careful attention to mental health implications for humanity as a whole.

The ethical implications of discovering extraterrestrial life are particularly complex and far-reaching. The question of whether we should attempt to contact extraterrestrial civilizations, and how we should behave if we do encounter them, are matters of significant debate. Should we adhere to a "prime directive" similar to that found in science fiction, refraining from interference to allow other life forms to develop naturally? Or should we actively engage with them, potentially sharing our knowledge and technology? These decisions have potentially profound consequences for both humanity and any extraterrestrial life we encounter. We must develop ethical guidelines that ensure that our interactions are conducted with respect, consideration, and a commitment to the long-term well-being of all life forms involved. This framework requires international cooperation and a careful consideration of potential risks and benefits.

The search for extraterrestrial life is therefore not merely a scientific endeavor; it is a journey of self-discovery, a quest to understand our place in the vastness of the cosmos. The philosophical, existential, and ethical implications are intertwined, inseparable from the scientific challenges. The discovery of life beyond Earth would not just add another chapter to our scientific understanding; it would rewrite the entire book of human history, forcing us to confront our deepest assumptions about ourselves and our place in the universe. The future holds immense possibilities, both exciting and daunting, and it is through careful consideration of all these implications that we can best prepare ourselves for the discoveries that lie ahead. The universe awaits, and our response to what we find will define humanity's future.

Looking back at the journey we've undertaken, from the microbial ecosystems thriving in the most extreme environments on Earth to the tantalizing possibilities of exoplanets orbiting distant suns, a profound sense of awe and wonder inevitably emerges. We've explored the diverse tapestry of life on our own planet, highlighting its resilience, adaptability, and astonishing capacity to flourish in a myriad of conditions. This biological diversity serves as a powerful testament to life's inherent drive to exist and evolve, suggesting that the conditions for life's emergence might be far more common than we once imagined. The sheer number of stars and planets in the observable universe, each with its own unique story to tell, underscores the immense potential for life beyond Earth. This potential is not merely a scientific hypothesis; it is a profound and compelling possibility that deserves serious consideration and continued exploration.

The sheer scale of the universe, however, also emphasizes the humbling reality of our current limitations. Our understanding of the cosmos is still nascent. Despite the advancements in telescopes, space probes, and data analysis techniques, vast swathes of the universe remain shrouded in mystery. We are only beginning to scratch the surface of understanding the conditions necessary for life to arise, the processes that drive its evolution, and the diversity of forms it might take. Even the definition of "life" itself continues to be refined as we uncover more about extremophiles and unusual biochemical processes on Earth. The possibility of extraterrestrial life that differs drastically from anything we know—life based on entirely different chemistries, with unconventional forms of energy utilization or reproduction—remains a very real and exciting possibility. Imagining such life forces us to expand our definitions, embrace the unknown, and acknowledge the limits of our current scientific paradigms.

The search for extraterrestrial life isn't just about finding other organisms; it's about finding ourselves, in a broader cosmic context. It's a quest for self-understanding, a profound reflection on our own origins and our place within the grand scheme of the universe. The question of whether we are alone, of whether life is a unique event confined to Earth or a common phenomenon spreading throughout the cosmos, is a question that cuts to the very core of our existence. The answer, whatever it may be, will reshape our understanding of humanity's significance, its responsibilities, and its future trajectory. It could redefine our relationship not only with the cosmos but also with ourselves, prompting profound philosophical and ethical considerations about our own actions and our place in the web of life.

One aspect often overlooked in the discussions surrounding extraterrestrial life is the potential for different forms of intelligence. Our current understanding of intelligence is largely based on our own human experience. We tend to associate intelligence with complex tool use, abstract thought, and advanced communication systems. However, extraterrestrial life might display entirely different forms of intelligence, forms that we might not even recognize as intelligent. Consider, for example, a highly interconnected network of organisms that communicate through subtle chemical signals, or a species that exists in a dimension or plane of reality beyond our current comprehension. The very concept of intelligence itself may require a broader definition, one that embraces the unexpected and transcends our anthropocentric biases.

This also highlights the limitations of our current methods of searching for extraterrestrial life. We often focus on signs of life that mirror our own – searching for water, oxygen, or organic molecules. While these are important indicators, they might not be universal markers of life. Extraterrestrial life might exist in environments and forms that defy our current expectations, making detection incredibly challenging. We must consider the possibility of life that is fundamentally different, life that might be so alien that it goes entirely undetected by our current technologies and search strategies. This necessitates a shift in our approach, a broadening of our perspective and a deeper understanding of the fundamental principles governing the emergence and evolution of life. This may involve the development of novel technologies and detection methods, moving beyond the limitations of our current understanding.

Furthermore, the possibility of encountering extraterrestrial life raises profound ethical questions that need careful consideration. Should we actively seek contact, or should we adhere to a “prime directive” of non-interference? What are our ethical responsibilities towards any extraterrestrial life we discover? How do we ensure the safety and well-being of both humanity and any potential extraterrestrial civilizations? The answers to these questions are far from straightforward. They require a global dialogue, involving scientists, ethicists, policymakers, and the public, to establish a framework for responsible and ethical interactions with extraterrestrial life. This ethical framework is not merely a hypothetical concern; it is a crucial aspect of planning for a future where the discovery of extraterrestrial life is not a matter of "if" but "when."

Beyond the immediate ethical implications, the long-term consequences of discovering extraterrestrial life are virtually impossible to fully anticipate. It could fundamentally alter our social structures, political systems, religious beliefs, and economic models. The discovery could usher in an era of unprecedented technological advancements, leading to breakthroughs in medicine, energy, and communication. However, it also carries the potential for conflict, competition, and the exploitation of extraterrestrial resources. Navigating this complex landscape will require careful planning, international cooperation, and a commitment to responsible innovation. The future of humanity, in a universe potentially teeming with life, hinges on our ability to make informed, ethical, and forward-thinking decisions.

Finally, we must acknowledge the immense and awe-inspiring mystery that lies ahead. The universe is vastly larger and far more complex than we can currently comprehend. Our current knowledge is just a tiny fraction of the vast expanse of the unknown. The challenges ahead are immense, but the potential rewards are even greater. The continued exploration of the universe and the relentless pursuit of answers to fundamental questions about life's origin and prevalence are not merely scientific endeavors but essential components of humanity's ongoing quest for knowledge and understanding. The mystery, the vastness of the unknown, should not deter us; instead, it should inspire and motivate us to delve deeper, to push the boundaries of our understanding, and to continue our journey into the incredible universe that surrounds us. The quest for extraterrestrial life is, in essence, a testament to humanity's enduring curiosity, our relentless pursuit of knowledge, and our profound desire to understand our place within this grand cosmic tapestry. The journey continues. The universe awaits. And the mysteries it holds promise to redefine our very understanding of existence itself.

www.ingramcontent.com/pod-product-compliance
Lightning Source LLC
LaVergne TN
LVHW040517210926
848480LV00007BA/1489
9798349368448